D1403826

CURRENT TOPICS IN

Cellular Regulation

Volume 7

Contributors to Volume 7

BOB B. BUCHANAN
EARL W. DAVIE
IRVING B. FRITZ
OSAMU HAYAISHI
TASUKU HONJO
NOBUHIKO KATUNUMA
EDWARD P. KIRBY
PETER SCHÜRMANN
N. E. TOLBERT

CURRENT TOPICS IN
Cellular Regulation

edited by

Bernard L. Horecker · Earl R. Stadtman

Albert Einstein College of Medicine
Bronx, New York

National Institutes of Health
Bethesda, Maryland

Volume 7

1973

ACADEMIC PRESS New York and London

A Subsidiary of Harcourt Brace Jovanovich, Publishers

ACADEMIC PRESS, INC.
111 Fifth Avenue, New York, New York 10003

United Kingdom Edition published by
ACADEMIC PRESS, INC. (LONDON) LTD.
24/28 Oval Road, London NW1

LIBRARY OF CONGRESS CATALOG CARD NUMBER: 72-84153

PRINTED IN THE UNITED STATES OF AMERICA

Contents

Ribulose 1,5-Diphosphate Carboxylase: A Regulatory Enzyme in the Photosynthetic Assimilation of Carbon Dioxide

BOB B. BUCHANAN AND PETER SCHÜRMANN

Glycolate Biosynthesis

N. E. TOLBERT

Molecular Mechanisms in Blood Coagulation

EARL W. DAVIE AND EDWARD P. KIRBY

Enzymatic ADP-Ribosylation of Proteins and Regulation of Cellular Activity

TASUKU HONJO AND OSAMU HAYAISHI

Selected Topics on the Biochemistry of Spermatogenesis

IRVING B. FRITZ

Enzyme Degradation and Its Regulation by Group-Specific Proteases in Various Organs of Rats

NOBUHIKO KATUNUMA

List of Contributors

Numbers in parentheses indicate the pages on which the authors' contributions begin.

BOB B. BUCHANAN (1), *Department of Cell Physiology, University of California, Berkeley, California*

EARL W. DAVIE (51), *Department of Biochemistry, University of Washington, Seattle, Washington*

IRVING B. FRITZ (129), *Banting and Best Department of Medical Research, University of Toronto, Toronto, Ontario, Canada*

OSAMU HAYAISHI (87), *Department of Medical Chemistry, Kyoto University Faculty of Medicine, Kyoto, Japan*

TASUKU HONJO* (87), *Department of Embryology, Carnegie Institution of Washington, Baltimore, Maryland*

NOBUHIKO KATUNUMA (175), *Department of Enzyme Chemistry, Institute for Enzyme Research, School of Medicine, Tokushima University, Tokushima, Japan*

EDWARD P. KIRBY (51), *Department of Biochemistry, Temple University School of Medicine, Philadelphia, Pennsylvania*

PETER SCHÜRMANN (1), *Department of Cell Physiology, University of California, Berkeley, California*

N. E. TOLBERT (21), *Department of Biochemistry, Michigan State University, East Lansing, Michigan*

Present address: Laboratory of Molecular Genetics, National Institute of Child Health and Human Development, National Institute of Health, Bethesda, Maryland.

Preface

Recent years have witnessed rapid advances in our knowledge of the basic mechanisms involved in the regulation of diverse cellular activities such as intermediary metabolism, the transfer of genetic information, membrane permeability, and cellular differentiation and other organ functions. Information gained from the detailed analyses of a large number of isolated enzyme systems, together with results derived from physiological investigations of metabolic processes *in vivo*, constitutes an ever-increasing body of knowledge from which important generalized concepts and basic principles of cellular regulation are beginning to emerge. However, so rapid are the present advances in the general area of cellular regulation and so diverse are the disciplines involved, that it has become a formidable task for even the expert in a specialized area to keep abreast of the progress in his field. This series of volumes is concerned with such recent developments in various areas of cellular regulation. We do not intend that it will consist of comprehensive annual reviews of the literature. We hope rather that it will constitute a medium which will, on the one hand, provide contributing authors with an opportunity to summarize progress in specialized areas of study that have undergone substantial developments and, on the other hand, serve as a forum for the enunciation of general principles and for the formulation of provocative theories and novel concepts. To this end editorial review of individual contributions will be concerned primarily with the clarity of presentation and conformity to publication policies. It is hoped in this manner to bring together current knowledge of various aspects of cellular regulation so as both to enlighten the uninformed and to provide a base of knowledge for those engaged in research in this subject.

BERNARD L. HORECKER
EARL R. STADTMAN

Preface to Volume 7

This volume explores a number of recently discovered areas and mechanisms of cellular regulation. The first two chapters take up the problem of regulation of the primary step in the incorporation of CO_2 in photosynthesis. Two puzzling earlier observations may now be explained—one is the apparent low affinity of purified RuDP carboxylase for CO_2 and the other the photoproduction of glycolic acid and the related effects of O_2 on CO_2 fixation. RuDP carboxylase now appears to be a site of regulation of photosynthesis. It may also be the first member of a novel class of oxygenase.

Other chapters are concerned with the mechanism of blood clotting and the events that may trigger this process, with the inhibition of protein synthesis by diphtheria toxin and the role of NAD and ADP ribosylation, with mechanisms of spermatogenesis and their control, and with the possible role of a novel type of protease, specific for pyridoxal phosphate enzymes, in the regulation of the levels of these enzymes in mammalian cells.

BERNARD L. HORECKER
EARL R. STADTMAN

Contents of Previous Volumes

Ribulose 1,5-Diphosphate Carboxylase: A Regulatory Enzyme in the Photosynthetic Assimilation of Carbon Dioxide

Bob B. Buchanan
and
Peter Schürmann

Department of Cell Physiology
University of California
Berkeley, California

I. Introduction

Interest in ribulose 1,5-diphosphate* (RuDP) was first aroused in 1951 when Benson (*1*) isolated and identified as a photosynthetic product this five-carbon sugar diphosphate, which was previously unknown in biological materials. This finding had great impact on future research in photosynthesis and led in 1954 to the proposal

---

* Abbreviations used are: RuDP, ribulose 1,5-diphosphate; PGA, 3-phosphoglyceric acid; F6P, fructose 6-phosphate; FDP, fructose 1,6-diphosphate; FDPase, fructose-1,6-diphosphatase.

1

by Calvin, Benson, and collaborators of the reductive pentose phosphate cycle (2) which includes as the sole step of CO_2 incorporation the carboxylation of RuDP to yield 3-phosphoglycerate (PGA). Evidence at that time for the postulated carboxylation of RuDP consisted basically of two findings on the kinetics of photosynthetic CO_2 assimilation in whole cells: (i) the identification of PGA as the first chemically defined product (3); and (ii) the formation of PGA from sugar diphosphates in light–dark transition experiments (4).

The carboxylation reaction was supported by simultaneous reports (submitted within 1 week of each other) from the laboratories of Calvin (5) and Horecker (6) that extracts of photosynthetic cells incubated with RuDP (5) or ribose 5-phosphate (6) as substrate catalyzed an incorporation of $^{14}CO_2$ into the carboxyl group of PGA. More extensive biochemical evidence for the reaction was described in the now classical papers from the laboratories of Horecker (7), Ochoa (8), and Racker (9) on the purification and properties of the enzyme ribulose 1,5-diphosphate carboxylase (RuDP carboxylase). The carboxylase was shown to be a Mg^{2+}-dependent enzyme distinct from the associated enzyme of the carbon reduction cycle, phosphoribulokinase and phosphoriboisomerase [both of which were also first purified in Horecker's laboratory (10)]. The stoichiometry of the carboxylation reaction, Eq. (1), was established with the purified enzyme (7, 8).

$$\text{Ribulose 1,5-diphosphate} + CO_2 + H_2O \xrightarrow{\text{Mg}^{2+}} \text{two 3-phosphoglycerate} \quad (1)$$

One property of RuDP carboxylase which appears to be in conflict with its proposed role in photosynthesis is a requirement for a CO_2 concentration about 100-fold greater than that normally present in air (7, 9, 11). The low affinity of RuDP carboxylase for CO_2 has remained a puzzling feature of the carboxylation reaction (cf. 11–14), and it gave rise in some quarters to doubts about the validity of the proposed carbon reduction cycle; for example, it was one reason given by the late Otto Warburg for rejecting the cycle (15, 16).

It is against this background that the affinity of the carboxylase for CO_2 was reexamined in our laboratory. Following a brief description of the properties of RuDP carboxylase, the results of this investigation will be summarized (17, 18). Certain new aspects of this work have been published in more detail elsewhere (19).

II. Properties of Ribulose 1,5-Diphosphate Carboxylase

A. General Characteristics

Since its initial isolation, RuDP carboxylase has continued to attract wide interest not only because of its key role in photosynthesis but also because of its unique features. First, the carboxylase is present at very high concentrations, accounting for up to 50% of the total soluble protein of leaves (11). The reason for such a large amount of RuDP carboxylase is not known; one early proposal (7), which still seems plausible, is that high concentrations of the enzyme are necessary to compensate for its low catalytic activity (see below).

It is of historical interest to note that, because of its high concentration, Wildman and Bonner (20) were able electrophoretically to identify RuDP carboxylase (called "fraction I protein") and to purify it 7 years before the RuDP carboxylation reaction was discovered in 1954 (5, 6). The similarities between the carboxylase and fraction I protein were noted early (21), but nearly a quarter of a century elapsed before the two proteins were conclusively shown to be identical (22).

Renewed interest in RuDP carboxylase was aroused recently by the finding (23) of a second enzymatic activity (see Tolbert, this volume). When assayed under conditions that prevent carboxylation of RuDP (high O_2 tension, no CO_2), the carboxylase behaves as an oxidase and oxidatively cleaves RuDP at the C-2 position to phosphoglycolate and PGA. This oxidase activity appears to play an important role in the synthesis of glycolate, but it is unrelated per se to the photosynthetic assimilation of CO_2.

On the basis of its carboxylase activity, RuDP carboxylase has been purified to homogeneity from algae and higher plants (11) and was recently crystallized from tobacco leaves (22). Algal and higher plant RuDP carboxylases have molecular weights of about

560,000 ($S_{20,w}$ = 18.5) and, on treatment with sodium dodecyl sulfate, urea, or guanidine-HCl, dissociate into subunits of molecular weights 54,000–60,000 and 12,000–16,000 (*24–32*).

Of particular importance to the present discussion are the kinetic constants of RuDP carboxylase. At saturating levels of Mg^{2+} (K_m = 1 × 10^{-3} M) (*33*) the spinach leaf carboxylase shows a strong affinity for one of the substrates, RuDP (K_m = 1 to 2.5 × 10^{-4} M) (*7, 33*) and a weak affinity for the other, bicarbonate (K_m = 1 to 2 × 10^{-2} M) (*7, 9, 33, 34*). The sluggish catalytic activity of the enzyme is reflected in its relatively low turnover rate of 1340 (micromoles of bicarbonate fixed per mole of enzyme per minute) (*33*). Some of the properties of the spinach leaf RuDP carboxylase, which in general are valid for other plant carboxylases (*13, 32, 34*), are summarized in Table I.

The important finding by Cooper *et al.* (*35*) that CO_2, rather than bicarbonate, is the carbon species which serves as the substrate for RuDP carboxylase merits mention. Correction of the Michaelis constant to the concentration of the active species, CO_2, results in a K_m of about 4.5 × 10^{-4} M, which corresponds approximately to the total "CO_2" content (CO_2 + HCO_3^- + H_2CO_3) of a solution (buffered at pH 7.9) in equilibrium with air (*32*). However, at pH 7.9, gaseous CO_2 represents less than 1% of the total carbon and in the absence of a mechanism of activation is incapable of supporting an appreciable rate of CO_2 fixation by the isolated carboxylase. In the discussion below, the term bicarbonate is used to represent the total "CO_2" content (CO_2 + HCO_3^- + H_2CO_3). No calculation has been made to convert the bicarbonate concentrations to the concentration of CO_2.

TABLE I

PROPERTIES OF RuDP CARBOXYLASE FROM SPINACH LEAVES[a]

Molecular weight	560,000
Number of subunits (nonidentical)	2
K_m, HCO_3^-	1 to 2 × 10^{-2} M
K_m, RuDP	1 to 2.5 × 10^{-4} M
K_m, Mg^{2+}	1 × 10^{-3} M
Turnover rate (μmoles CO_2 fixed/mole enzyme/min)	1340

[a] Appropriate references are given in the text.

B. Comparison of the Bicarbonate Requirement for Complete Photosynthesis by Isolated Chloroplasts and for Carboxylation of Ribulose 1,5-Diphosphate by Chloroplast Extract

The discovery that isolated chloroplasts photoassimilate bicarbonate to the level of carbohydrate (*36, 37*) extended more recently by findings that this reaction can be made to proceed at high rates (up to 60% those of leaves) (*38–41*) permits an estimation *in vitro* of the concentration of bicarbonate needed to sustain a maximal rate of photosynthesis. Isolated chloroplasts have a further advantage in that (after breaking and removing the chlorophyllous fragments) the RuDP carboxylase component of the soluble chloroplast protein fraction ("chloroplast extract") (*42–47*) may be used to determine in the same preparation the concentration of bicarbonate required for carboxylation of RuDP. Figure 1 shows a

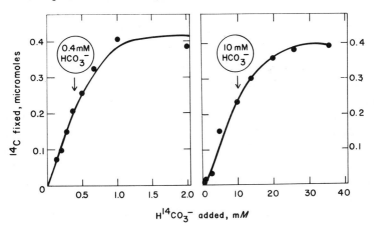

Fig. 1. Comparison of the bicarbonate requirements for complete photosynthesis by isolated chloroplasts and for carboxylation of ribulose 1,5-diphosphate (RuDP) by chloroplast extract. Complete photosynthesis by isolated chloroplasts (left panel) was determined as described previously (*53*). RuDP carboxylation by chloroplast extract (right panel) (*62*) was determined with the assay described for pure RuDP carboxylase in Table II, except that enzyme modifiers were omitted, the reaction time was 6 minutes, and chloroplast extract (equivalent to 75 μg of chlorophyll) replaced the pure carboxylase.

marked difference in the bicarbonate requirement of the intact and broken chloroplast preparations (both at pH 7.5); the half-maximal rate of CO_2 assimilation by intact chloroplasts occurred at 0.4 mM bicarbonate (optimal level, 1.0 mM), whereas the half-maximal carboxylation rate of RuDP by chloroplast extract occurred at 10 mM bicarbonate (optimal level, 25 mM). These results and those of others (13, 40, 48) show that isolated chloroplasts have a much greater affinity for bicarbonate than does the enzyme RuDP carboxylase. It would therefore appear that intact chloroplasts possess a mechanism (lost on breakage of the chloroplast membrane) for activating RuDP carboxylase at a low concentration of bicarbonate. Several investigators (49–52) have proposed that the carboxylase is activated by light, but until recently a specific regulatory mechanism was unknown.

III. Regulation of Ribulose 1,5-Diphosphate Carboxylase

A. Effect of Carbon Intermediates and Related Constituents of Chloroplasts

Our search for a regulatory mechanism in the carboxylation reaction centered first on the soluble products which isolated chloroplasts accumulate during photosynthesis (Fig. 2). About 90% of the total bicarbonate-^{14}C fixed into soluble compounds after 12-min photosynthesis is recovered in PGA and sugar mono- and diphosphates, most of which are intermediates of the reductive pentose phosphate cycle (53). When these intermediates of the cycle were tested (Table II) at the bicarbonate concentration optimal for photosynthesis by isolated chloroplasts, they were found to include: (i) intermediates which activated the enzyme (notably F6P and, to a lesser extent, ribulose 5-phosphate, xylulose 5-phosphate, erythrose 4-phosphate, and ribose 5-phosphate); (ii) intermediates which had essentially no effect (dihydroxyacetone phosphate, sedoheptulose 7-phosphate, and PGA); and (iii) one intermediate, FDP, which inhibited the enzyme. Intermediates in starch synthesis either activated slightly (glucose 6-phosphate and ADP-glucose) or inhibited (glucose 1,6-diphosphate and glucose 1-phosphate) the carboxylase. ADP and ATP stimulated to a small degree; AMP, cyclic

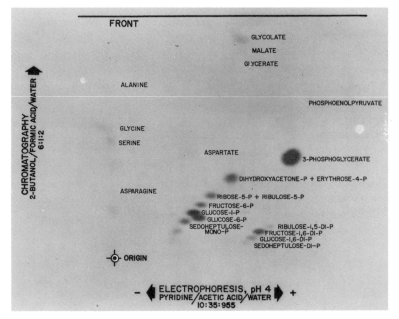

FIG. 2. Radioautograph of bicarbonate-^{14}C assimilation by isolated chloroplasts (53).

AMP, and fructose had no effect; fructose 1-phosphate and phosphoenolpyruvate inhibited.

B. Factors Influencing Activation by Fructose 6-Phosphate and Deactivation by Fructose 1,6-Diphosphate

The stimulation of RuDP carboxylase activity by F6P and its inhibition by FDP suggested that these two intermediates of the carbon cycle could regulate carboxylase activity during photosynthesis, with F6P activating and FDP deactivating the enzyme. The other stimulatory intermediates (ribulose 5-phosphate, xylulose 5-phosphate, erythrose 4-phosphate, and ribose 5-phosphate) are present at lower concentration in chloroplasts (53) and stimulated the carboxylation reaction less than did F6P. Of the inhibitory compounds, FDP was considered to be of greatest potential importance because of its key position in the reductive pentose phosphate cycle.

As shown in Fig. 3, RuDP carboxylase that was activated by

TABLE II

EFFECT OF INTERMEDIATES OF THE REDUCTIVE PENTOSE PHOSPHATE
CYCLE AND RELATED COMPOUNDS ON ACTIVITY OF PURE RuDP
CARBOXYLASE FROM SPINACH LEAVES[a]

Stimulatory compounds	Carboxylase activity (% of control)	Ineffective and inhibitory compounds	Carboxylase activity (% of control)
Carbon cycle intermediates		Carbon cycle intermediates	
None (Control)	100	None (Control)	100
F6P	660	Dihydroxyacetone	
Ribulose 5-phosphate	460	phosphate	133
Xylulose 5-phosphate	370	Sedoheptulose	
Erythrose 4-phosphate	305	7-phosphate	120
Ribose 5-phosphate	220	PGA	94
Related compounds		FDP	61
ADP	223	Related compounds	
ADP-Glucose	200	AMP	118
ATP	150	Cyclic (3′,5′)-AMP	111
Glucose 6-phosphate	147	Fructose	101
		Phosphoenolpyruvate	68
		Fructose 1-phosphate	57
		Glucose 1,6-diphosphate	51
		Glucose 1-phosphate	51

[a] The reaction mixture contained the following (added in the order indicated): Tricine buffer, pH 7.5, 200 mM; EDTA, 0.06 mM; MgCl$_2$, 1 mM; reduced glutathione, 5 mM; RuDP carboxylase [purified to homogeneity by a procedure based on that of Paulsen and Lane (33)], 20 μg; each of the indicated compounds, 0.5 mM; NaH^{14}CO$_3$ (5 × 10^5 cpm/μmole), 1 mM; and RuDP, 0.3 mM. Volume, 0.5 ml; temperature, 25°; reaction time, 10 minutes. The reaction was carried out in scintillation vials and was stopped with 0.05 ml of 6 N HCl; samples were evaporated to dryness, and the ^{14}C-PGA formed was determined in a scintillation counter. The control treatment gave 510 cpm of PGA formed.

concentrations of F6P ranging from 0.04 to 0.4 mM was deactivated by corresponding concentrations of FDP. The effect of several factors on the activation and deactivation of RuDP carboxylase under these conditions is described below.

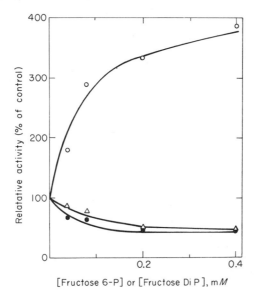

FIG. 3. Ribulose 1,5-diphosphate (RuDP) carboxylase: Activation by fructose 6-phosphate (F6P) and deactivation by fructose 1,6-diphosphate (FDP). Except for a reaction time of 3 minutes, using 50 μg of pure carboxylase, and varying concentrations of F6P and FDP, experimental conditions were as described for Table II. 3-Phosphoglyceric acid was identified as the reaction product in each of the treatments by thin-layer electrophoresis chromatography (*53*). ○——○, F6P; △——△, F6P + FDP; ●——●, FDP.

1. BICARBONATE CONCENTRATION

The concentration of bicarbonate greatly influenced the activation of RuDP carboxylase induced by F6P. At bicarbonate concentrations of 10 mM or less, F6P strikingly stimulated the carboxylase and thereby effected a shift in the bicarbonate concentration curve from the original sigmoidal type, seen previously by Sugiyama *et al.* (*51*) and Andrews and Hatch (*54*), to a hyperbolic type. At 20 mM or greater bicarbonate concentrations, F6P had little effect. Deactivation by FDP also depended on the bicarbonate concentration: at the lower bicarbonate concentrations, activity was reduced to the control value, whereas at higher bicarbonate concentrations activity was inhibited below the control value (cf. Ref. *55*).

Like other regulatory enzymes (*56, 57*), the untreated carboxylase

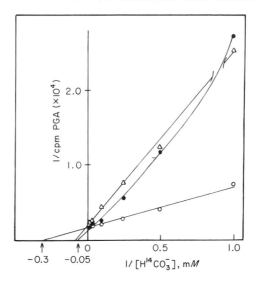

FIG. 4. Effect of fructose 6-phosphate (F6P) and fructose 1,6-diphosphate (FDP) on the K_m of ribulose 1,5-diphosphate (RuDP) carboxylase for HCO_3^-. Except for varying the HCO_3^- concentration, experimental conditions were as described for Fig. 3. K_m, HCO_3^-: ●——●, control, $20.0 \times 10^{-3} M$; △——△, F6P + FDP, $12.5 \times 10^{-3} M$; ○——○, F6P, $3.3 \times 10^{-3} M$.

(control, Fig. 4) exhibited nonlinear kinetics. Activation by F6P changed the kinetics to the Michaelian (linear) type and lowered the K_m for bicarbonate by a factor of six (from $20 \times 10^{-3} M$ to $3.3 \times 10^{-3} M$). On addition of FDP, the K_m for bicarbonate increased 4-fold and, in certain experiments (which showed full restoration to the original kinetics), returned to the control value.

2. Mg²⁺ CONCENTRATION

The concentration of Mg^{2+} did not qualitatively affect activation by F6P but did change the magnitude of the activation; at a saturating concentration (20 mM Mg^{2+}), activation was half that at a limiting concentration (1 mM Mg^{2+}).

A determination of the kinetic constants for Mg^{2+} (Fig. 5) revealed that F6P lowered by a factor of five the K_m for Mg^{2+}. On deactivation, FDP effected a return of the K_m to about its initial value.

Recognition of the importance of Mg^{2+} in control of RuDP car-

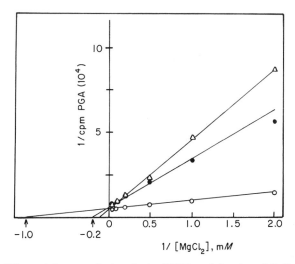

FIG. 5. Effect of fructose 6-phosphate (F6P) and fructose 1,6-diphosphate (FDP) on the K_m of ribulose 1,5-diphosphate (RuDP) carboxylase for Mg^{2+}. Except for varying the Mg^{2+} concentration, experimental conditions were as described for Fig. 3. K_m, Mg^{2+}: ●——●, control, $5 \times 10^{-3}\ M$; △——△, F6P + FDP, $10 \times 10^{-3}\ M$; ○——○, F6P, $1 \times 10^{-3}\ M$.

boxylase prompted us to reexamine the effect of bicarbonate at saturating Mg^{2+}. An up to 2-fold activation by F6P was observed at the lower bicarbonate concentrations, but activation disappeared at bicarbonate concentrations of 10 mM or greater. At saturating Mg^{2+}, F6P halved the K_m for bicarbonate (from $6.7 \times 10^{-3}\ M$ to $3.3 \times 10^{-3}\ M$) and promoted the characteristic shift from nonlinear to Michaelian kinetics (Fig. 6). Deactivation by FDP restored the original kinetics and affinity for bicarbonate.

The kinetic constants for RuDP were also determined at a saturating Mg^{2+} concentration. The K_m of the carboxylase for RuDP was $1 \times 10^{-4}\ M$ for the control, $5 \times 10^{-4}\ M$ for the enzyme activated by F6P, and $4 \times 10^{-4}\ M$ for the enzyme deactivated by FDP. Thus, on activation by F6P, the affinity of RuDP carboxylase for RuDP appears to decrease in a manner only partly reversible by FDP.

3. pH

Activation of RuDP carboxylase by F6P was influenced not only by the concentration of bicarbonate and Mg^{2+} but also by pH.

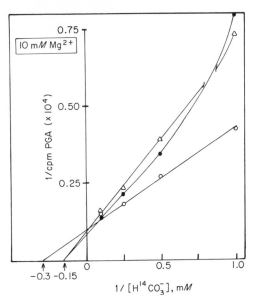

Fig. 6. Effect of fructose 6-phosphate (F6P) and fructose 1,6-diphosphate (FDP) on the K_m of ribulose 1,5-diphosphate (RuDP) carboxylase for HCO_3^- at a saturating concentration of Mg^{2+}. Except for varying the HCO_3^- concentration and adding 10 mM MgCl$_2$, experimental conditions were as described for Fig. 3. K_m, HCO_3^-: ●——●, control, $6.7 \times 10^{-3}\ M$; △——△, F6P + FDP, $6.7 \times 10^{-3}\ M$; ○——○, $3.3 \times 10^{-3}\ M$ F6P.

Total activity of the enzyme was highest at pH 7.5 but, as shown in Fig. 7, the effect of F6P was greatest at pH 7.0 (with 1 mM Mg^{2+}) or at pH 6.5 (with 20 mM Mg^{2+}). Deactivation by FDP occurred throughout the pH range tested. Under our conditions, there was no evidence of the reported (51, 58) shift of the pH optimum of the carboxylase to the acid region on increasing the concentration of Mg^{2+}.

To recapitulate, F6P activates RuDP carboxylase by increasing its affinity for both bicarbonate and Mg^{2+} (Table III). On activation, the carboxylase shows a shift, with respect to bicarbonate, from nonlinear to Michaelian kinetics. Increase in the affinity for bicarbonate is most striking at a limiting level of Mg^{2+} or at neutral or acidic pH. F6P also increases the affinity of the carboxylase for

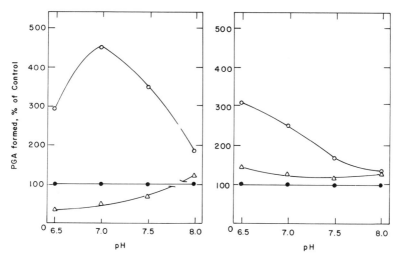

FIG. 7. Relative effect of fructose 6-phosphate (F6P) and fructose 1,6-diphosphate on ribulose 1,5-diphosphate (RuDP) carboxylase as a function of pH and [Mg^{2+}]. Except for varying the pH and Mg^{2+} concentration, experimental conditions were as described for Fig. 3. The 3-phosphoglyceric acid (PGA) formed at pH 7.5 in the presence of F6P was 11,000 cpm at 1 mM Mg^{2+} (left panel) and 22,000 cpm at 20 mM Mg^{2+} (right panel). ○——○, F6P; ●——●, control; △——△, F6P + FDP.

TABLE III
KINETIC CONSTANTS OF RuDP CARBOXYLASE[a]

RuDP carboxylase treatment	K_m, HCO_3^- (1 mM Mg^{2+}) M	K_m, HCO_3^- (10 mM Mg^{2+}) M	K_m, Mg^{2+} (1 mM HCO_3^-) M	K_m, RuDP (10 mM Mg^{2+} 1 mM HCO_3^-) M
Control	20×10^{-3}	6.7×10^{-3}	5×10^{-3}	1×10^{-4}
Activated by F6P	3.3×10^{-3}	3.3×10^{-3}	1×10^{-3}	5×10^{-4}
Deactivated by FDP	12.5×10^{-3}	6.7×10^{-3}	10×10^{-3}	4×10^{-4}

[a] Constants were determined with 0.4 mM F6P and FDP added as indicated by using the assay described for Fig. 3.

Mg^{2+} and decreases its affinity for RuDP. FDP deactivates the carboxylase and effects, in general, a restoration of the original kinetic constants and the characteristic kinetics.

TABLE IV

ACTIVATION OF RuDP CARBOXYLASE BY 6-PHOSPHOGLUCONATE
AND DEACTIVATION BY FRUCTOSE DIPHOSPHATE[a]

Modifier added	3-Phosphoglyceric acid formed (% of control)
None	100
6-Phosphogluconate	402
6-Phosphogluconate + FDP	54
F6P	285
F6P + FDP	55

[a] F6P (0.4 mM) and 6-phosphogluconate (0.2 mM) were added as indicated. FDP concentration was 0.4 mM when added with F6P and 0.2 mM when added with 6-phosphogluconate. Other experimental conditions were as described for Fig. 3.

C. Activation by 6-Phosphogluconate

6-Phosphogluconate has recently been implicated in the control of RuDP carboxylase by virtue of its inhibition of carboxylase activity (59, 60). In testing the effect of 6-phosphogluconate under our assay conditions, we found not the expected inhibition but an even greater activation than with F6P (Table IV). Like F6P, activation by 6-phosphogluconate was reversible by FDP and was restricted to low bicarbonate concentrations; at saturating bicarbonate, 6-phosphogluconate inhibited markedly (cf. 59, 60).

On a molar basis, 6-phosphogluconate was more effective than F6P in activating RuDP carboxylase, but otherwise the two compounds were similar. 6-Phosphogluconate was most effective at low concentrations of Mg^{2+} and lowered the Michaelis constant for both bicarbonate and Mg^{2+} (see tabulation). At present it is not known

	K_m, Mg^{2+} M	K_m, HCO_3^- (1 mM Mg^{2+}) M	K_m, HCO_3^- (10 mM Mg^{2+}) M
Control	4.4×10^{-3}	2.0×10^{-2}	1.0×10^{-2}
+6-Phosphogluconate	0.7×10^{-3}	0.25×10^{-2}	0.25×10^{-2}

whether the activation by 6-phosphogluconate is of physiological significance in chloroplasts.

IV. RuDP Carboxylase and Regulation of Photosynthesis

A remaining component in the regulation of photosynthetic CO_2 assimilation is the mechanism for forming the F6P and FDP needed to regulate RuDP carboxylase. In CO_2 assimilation by chloroplasts where F6P and FDP are intermediates of the reductive pentose phosphate cycle, F6P is formed hydrolytically from FDP by fructose 1,6-diphosphatase (FDPase)—an enzyme which itself is under regulatory control (49, 50, 61–63). The mechanism of FDPase control in chloroplasts appears, however, to be different from that in nonphotosynthetic cells. Chloroplast FDPase (purified to homogeneity) (61, 63) differs from its counterpart in nonphotosynthetic cells (64–66) in being insensitive to AMP (61, 63, 67). The regulation of the chloroplast FDPase system depends on ferredoxin and a protein factor (62, 63). In the presence of the protein factor, reduced ferredoxin activates the FDPase enzyme according to Eqs. (2) and (3) and thereby enhances the formation of F6P from FDP (Eq. 4).

$$4 \text{ Ferredoxin}_{\text{oxidized}} + 2 \text{ H}_2\text{O} \xrightarrow[\text{light}]{\text{chloroplasts}} 4 \text{ ferredoxin}_{\text{reduced}} + \text{O}_2 \quad (2)$$

$$\text{FDPase}_{\text{inactive}} + \text{ferredoxin}_{\text{reduced}} \xrightarrow[\text{factor}]{\text{protein}}$$
$$\text{FDPase}_{\text{active}} + \text{ferredoxin}_{\text{oxidized}} \quad (3)$$

$$\text{FDP} + \text{H}_2\text{O} \xrightarrow[\text{Mg}^{2+}]{\text{active FDPase}} \text{F6P} + \text{P}_i \quad (4)$$

Through its control of the levels of F6P and FDP in chloroplasts, the ferredoxin-FDPase system may provide a mechanism for regulating the activity of RuDP carboxylase (Fig. 8). RuDP carboxylase, deactivated by FDP in the dark, would be activated by F6P formed via the ferredoxin-FDPase system in the light. If such a regulatory system were operative, it should be possible to show in isolated, intact chloroplasts: (i) activation of FDPase by light; (ii) activation of RuDP carboxylase by light; and (iii) activation of RuDP carboxylase by F6P. Tables V and VI present evidence for each of these expected effects (cf. refs. 48, 49).

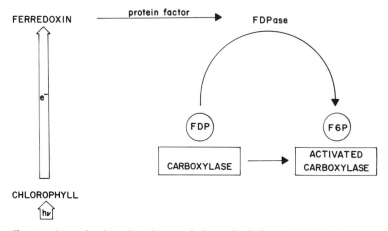

Fig. 8. A mechanism for the regulation of ribulose 1,5-diphosphate carboxylase in photosynthesis. FDP, fructose 1,6-diphosphate, F6P, fructose
6-phosphate.

TABLE V
Activation by Light of the
FDPase of Intact
Chloroplasts (63)

	P_i released (μmole)
Dark, Control	0.5
Light, Control	0.9
Light, minus FDP	0.0

V. Concluding Remarks

The activity of RuDP carboxylase in plant photosynthesis appears
to be controlled by the relative concentrations of F6P and FDP;
these, in turn, are governed by the light-dependent, ferredoxin-activated FDPase system. The immediate effect of light is to enhance
the formation of F6P from FDP, which then activates RuDP carboxylase. The FDP accumulated in the dark deactivates the
carboxylase.

An additional control point discovered by Preiss and collaborators

TABLE VI
EQUIVALENCE OF LIGHT AND F6P
IN ACTIVATION OF RIBULOSE
1,5-DIPHOSPHATE CARBOXYLASE
OF INTACT CHLOROPLASTS

Treatment	Relative carboxylase activity
Dark	100
Light	152
Dark (+F6P)	146

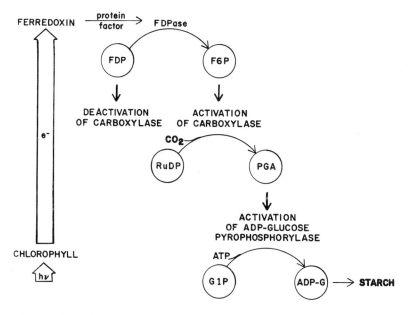

FIG. 9. A regulatory mechanism for the photosynthetic formation of starch. FDP, fructose 1,6-diphosphate; F6P, fructose 6-phosphate; RuDP, ribulose 1,5-diphosphate; PGA, 3-phosphoglyceric acid; G1P, glucose 1-phosphate; ADP, ATP, adenosine di- and triphosphates.

(*68, 69*) operates at the level of starch synthesis in chloroplasts. The enzyme that forms ADP-glucose for starch synthesis, ADP-glucose pyrophosphorylase (Eq. 5) was found to be a regulatory enzyme that is activated by PGA.

$$\text{ATP} + \text{glucose 1-phosphate} \xrightarrow[\text{pyrophosphorylase}]{\text{ADP-glucose}}$$
$$\text{ADP-glucose} + \text{pyrophosphate} \quad (5)$$

Through its dependency on PGA, the regulation of ADP-glucose pyrophosphorylase activity may be linked to the ferredoxin-dependent FDPase system (Fig. 9). In such a mechanism, the ferredoxin-activated FDPase is the initial regulatory reaction for the total synthesis of starch from CO_2. RuDP carboxylase and ADP-glucose pyrophosphorylase would represent additional control points that are governed, respectively, by the concentration of F6P and FDP and of PGA. Through its requirement for reduced ferredoxin, such a regulatory mechanism would be dependent on light.

ACKNOWLEDGMENT

We gratefully acknowledge the advice and support of Professor D. I. Arnon. Figures 1,4–7, and 9 are reproduced from *J. Biol. Chem.* **248**, 4956 (1973); Fig. 2 is from *Biochim. Biophys. Acta* **267**, 117 (1972).

REFERENCES

1. Benson, A. A., *J. Amer. Chem. Soc.* **73**, 2971 (1951).
2. Bassham, J. A., Benson, A. A., Kay, L. D., Harris, A. Z., Wilson, A. T., and Calvin, M., *J. Amer. Chem. Soc.* **76**, 1760 (1954).
3. Benson, A. A., Bassham, J. A., Calvin, M., Goodale, T. C., Hass, V. A., and Stepka, W., *J. Amer. Chem. Soc.* **72**, 1710 (1950).
4. Calvin, M., and Massini, P., *Experientia* **8**, 445 (1952).
5. Quayle, J. R., Fuller, R. C., Benson, A. A., and Calvin, M., *J. Amer. Chem. Soc.* **76**, 3610 (1954).
6. Weissbach, A., Smyrniotis, P. Z., and Horecker, B. L., *J. Amer. Chem. Soc.* **76**, 3611 (1954).
7. Weissbach, A., Horecker, B. L., and Hurwitz, J., *J. Biol. Chem.* **218**, 795 (1956).
8. Jakoby, W. B., Brummond, D. O., and Ochoa, S., *J. Biol. Chem.* **218**, 811 (1956).
9. Racker, E., *Arch. Biochem. Biophys.* **69**, 300 (1957).
10. Hurwitz, J., Weissbach, A., Horecker, B. L., and Smyrniotis, P. Z., *J. Biol. Chem.* **218**, 769 (1956).

11. Kawashima, N., and Wildman, S. G., *Annu. Rev. Plant Physiol.* **21**, 325 (1970).
12. Stiller, M., *Annu. Rev. Plant Physiol.* **13**, 151 (1962).
13. Gibbs, M., Latzko, E., Everson, R. G., and Cockburn, W., *in* "Harvesting the Sun: Photosynthesis in Plant Life" (A. San Pietro, F. A. Greer, and T. J. Army, eds.), p. 111. Academic Press, New York, 1967.
14. Walker, D. A., and Crofts, A. R., *Annu. Rev. Biochem.* **39**, 389 (1970).
15. Warburg, O., Krippahl, G., Jetschmann, K., and Lehmann, A., *Z. Naturforsch. B* **18**, 837 (1963).
16. Warburg, O., and Krippahl, G., *Hoppe-Seyler's Z. Physiol. Chem.* **322**, 225 (1963).
17. Buchanan, B. B., and Schürmann, P., *Fed. Proc., Fed. Amer. Soc. Exp. Biol.* **31**, 261 (1972) (abstr.).
18. Buchanan, B. B., and Schürmann, P., *FEBS Let.* **23**, 157 (1972).
19. Buchanan, B. B., and Schürmann, P., *J. Biol. Chem.* **248**, 4956 (1973).
20. Wildman, S. G., and Bonner, J., *Arch. Biochem.* **14**, 381 (1947).
21. Donner, R. W., Kahn, A., and Wildman, S. G., *J. Biol. Chem.* **229**, 945 (1957).
22. Kawashima, N., and Wildman, S. G., *Biochim. Biophys. Acta* **229**, 240 (1971).
23. Bowes, G., Ogren, W. L., and Hageman, R. H., *Biochem. Biophys. Res. Commun.* **45**, 716 (1971).
24. Rutner, A. C., and Lane, M. D., *Biochem. Biophys. Res. Commun.* **28**, 531 (1967).
25. Sugiyama, T., and Akazawa, T., *J. Biochem. (Tokyo)* **62**, 474 (1967).
26. Moon, K. E., and Thompson, E. O. P., *Aust. J. Biol. Sci.* **22**, 463 (1969).
27. Kawashima, N., *Plant Cell Physiol.* **10**, 31 (1969).
28. Matsumoto, C., Sugiyama. T., Akazawa, T., and Miyachi, S., *Arch. Biochem. Biophys.* **135**, 282 (1969).
29. Sugiyama, T., and Akazawa, T., *Biochemistry* **9**, 4499 (1970).
30. Kawashima, N., and Wildman, S. G., *Biochim. Biophys. Acta* **299**, 749 (1971).
31. Rutner, A. C., *Biochem. Biophys. Res. Commun.* **39**, 923 (1970).
32. Siegel, M. I., Wishnick, M., and Lane, M. D., *in* "The Enzymes" (P. D. Boyer, ed.), 3rd ed., Vol. 6, p. 169. Academic Press, New York, 1972.
33. Paulsen, J. M., and Lane, M. D. *Biochemistry* **5**, 2350 (1966).
34. Kieras, F. J., and Haselkorn, R., *Plant Physiol.* **43**, 1264 (1968).
35. Cooper, T. G., Filmer, D., Wishnick, M., and Lane, M. D., *J. Biol. Chem.* **244**, 1081 (1969).
36. Arnon, D. I., Allen, M. B., and Whatley, F. R., *Nature (London)* **174**, 394 (1954).
37. Allen, M. B., Arnon, D. I., Capindale, J. B., Whatley, F. R., and Durham, L. J. *J. Amer. Chem. Soc.* **77**, 4149 (1955).
38. Walker, D. A., *Biochem. J.* **92**, 22C (1964).
39. Bucke, C., Walker, D. A., and Baldry, C. W., *Biochem. J.* **101**, 636 (1966).
40. Jensen, R. G., and Bassham, J. A., *Proc. Nat. Acad. Sci. U.S.* **56**, 1095 (1966).

41. Kalberer, P. P., Buchanan, B. B., and Arnon, D. I., *Proc. Nat. Acad. Sci. U.S.* **57**, 1542 (1967).
42. Whatley, F. R., Allen, M. B., Rosenberg, L. L., Capindale, J. B., and Arnon, D. I., *Biochim. Biophys. Acta* **20**, 462 (1956).
43. Trebst, A. V., Tsujimoto, H. Y., and Arnon, D. I., *Nature (London)* **182**, 351 (1958).
44. Losada, M., Trebst, A. V., and Arnon, D. I., *J. Biol. Chem.* **235**, 832 (1960).
45. Park, R. B., and Pon, N. G., *J. Mol. Biol.* **3**, 1 (1961).
46. Lyttleton, J. W., *Exp. Cell Res.* **26**, 312 (1962).
47. Trown, P. W., *Biochemistry* **4**, 908 (1965).
48. Jensen, R. G., *Biochim. Biophys. Acta* **234**, 360 (1971).
49. Pedersen, T. A., Kirk, M., and Bassham, J. A., *Physiol. Plant.* **19**, 219 (1966).
50. Jensen, R. G., and Bassham, J. A., *Biochim. Biophys. Acta* **153**, 227 (1968).
51. Sugiyama, T., Nakayama, N., and Akazawa, T., *Arch. Biochem. Biophys.* **126**, 737 (1968).
52. Wildner, G. F., and Criddle, R. S., *Biochem. Biophys. Res. Commun.* **37**, 952 (1969).
53. Schürmann, P., Buchanan, B. B., and Arnon, D. I., *Biochim. Biophys. Acta* **267**, 111 (1972).
54. Andrews, T. J., and Hatch, M. D., *Phytochemistry* **10**, 9 (1971).
55. Bowes, G., and Ogren, W. L., *J. Biol. Chem.* **247**, 2171 (1972).
56. Atkinson, D. E., *Annu. Rev. Biochem.* **35**, 85 (1966).
57. Stadtman, E. R., *Advan. Enzymol.* **28**, 41 (1966).
58. Bassham, J. A., Sharp, P., and Morris, I., *Biochim. Biophys. Acta* **153**, 989 (1968).
59. Tabita, F. R., and McFadden, B. A., *Biochem. Biophys. Res. Commun.* **48**, 1153 (1972).
60. Chu, D. K., and Bassham, J. A., *Plant Physiol.* **50**, 224 (1972).
61. Preiss, J., Biggs, M. L., and Greenberg, E., *J. Biol. Chem.* **242**, 2292 (1967).
62. Buchanan, B. B., Kalberer, P. P., and Arnon, D. I., *Biochem. Biophys. Res. Commun.* **29**, 74 (1967).
63. Buchanan, B. B., Schürmann, P., and Kalberer, P. P., *J. Biol. Chem.* **246** 5952 (1971).
64. Taketa, K., and Pogell, B. M., *Biochem. Biophys. Res. Commun.* **12**, 229 (1963).
65. Newsholme, E. A., *Biochem. J.* **89**, 38P (1963).
66. Mendicino, J., and Vasarhely, F., *J. Biol. Chem.* **238**, 3528 (1963).
67. Scala, J., Patrick, C., and Macbeth, G., *Arch. Biochem. Biophys.* **127**, 576 (1968).
68. Ghosh, H. P., and Preiss, J., *J. Biol. Chem.* **241**, 4491 (1966).
69. Preiss, J., and Kosuge, T., *Annu. Rev. Plant Physiol.* **21**, 433 (1970).

Glycolate Biosynthesis*

N. E. TOLBERT

*Department of Biochemistry
Michigan State University
East Lansing, Michigan*

I. Introduction

During photosynthesis large amounts of glycolate are formed in the chloroplasts and subsequently metabolized in the peroxisomes and mitochondria by the glycolate pathway. Photorespiration in

* Supported in part by NSF grant GB 32040X.

plants is a light-dependent uptake of O_2 and release of CO_2 during this biosynthesis and metabolism of glycolate. Gross rates of photosynthesis of about 70 mg of CO_2 fixed per square decimeter of leaf per hour are reduced by rates of CO_2 evolution or photorespiration of between 5 and 35 mg of CO_2 evolved, so that net photosynthesis may be lowered as much as 50%. The equilibrium established in a closed system between CO_2 uptake during photosynthesis and CO_2 loss during photorespiration is designated as the compensation point. This value is about 40–70 ppm of CO_2 for many C_3-plants in which the photosynthetic reductive pentose phosphate pathway for CO_2 fixation dominates, and it may be as high as 155 ppm for leaves of some trees. In crassulacean acid plants and in C_4-plants little CO_2 loss occurs because of an efficient CO_2 trapping and storage cycle involving oxaloacetate, malate, and aspartate, in addition to the photosynthetic carbon cycle. In these C_4-plants glycolate biosynthesis may be less, but it is not eliminated. Most of this review is concerned with glycolate biosynthesis by plants, whereas more details about its metabolism can be found elsewhere (80–82). Little is known about glycolate biosynthesis in animals, although this compound is present in blood and liver, and the kidney and liver peroxisomal glycolate oxidase is similar to that in leaf peroxisomes (20, 81).

Hypotheses concerning glycolate biosynthesis during photosynthetic CO_2 fixation should account for its formation in chloroplasts; the glycolate must be labeled uniformly with ^{14}C; O_2 uptake must occur; and $^{18}O_2$ must be incorporated only into the carboxyl group of the glycolate. In addition glycolate formation is greatly favored by low CO_2 concentrations and by high pH, and one mechanism should utilize the specific P-glycolate* phosphatase of the chloroplasts. These criteria are all fulfilled by the cleavage of RuDP* by RuDP oxygenase to P-glycolate and P-glycerate, whereas other hypotheses to be reviewed do not meet all of these physiological parameters. Criteria for glycolate formation in mammalian tissues and in non-green plant tissues have not been established.

---

* Abbreviations: RuDP, ribulose diphosphate; TPP-C_2, a thiamine pyrophosphate dihydroxyethyl complex; P-glycolate, phosphoglycolate; P-glycerate, phosphoglycerate.

II. ¹⁴C Tracer Experiments Establishing the Glycolate Pathway

Glycolate, glycine, and serine are labeled rapidly during $^{14}CO_2$ fixation by algae and plants (*8, 16*). Initially the glycolate is uniformly labeled (*15–17, 31, 32, 74, 103*), and so are the subsequent products, glycine and serine, in time periods so short (4–30 seconds) that the 3-P-glycerate from the photosynthetic carbon cycle remains predominantly carboxyl labeled (*15, 31, 62, 86*). Earlier investigations on photosynthesis in Calvin's group established that carbon atoms 1 and 2 of RuDP, as well as other hexose and heptulose phosphates, were also labeled at nearly uniform rates (*5*). Likewise addition of ribose-1-¹⁴C to tobacco seedlings (*29*) and 1-¹⁴C-labeled ribose-5-P to spinach chloroplasts produced small amounts of glycine-2-¹⁴C (*89, 101*). Thus it has long been proposed that glycolate arose from carbon atoms 1 and 2 of a sugar phosphate of the photosynthetic carbon cycle, after cleavage between atoms 2 and 3.

The glycolate pathway of metabolism in plants and algae as shown in Fig. 1 was established by feeding specifically labeled glycolate-¹⁴C, glycine-¹⁴C, and serine-¹⁴C and then measuring the distribution of the ¹⁴C among the carbon atoms in the products, glycine, serine, glycerate, and hexoses (*15, 31, 37, 62, 79, 86*). As one example, glycolate-2-¹⁴C in leaves is converted into serine-2,3-¹⁴C, glycerate-2,3-¹⁴C, and hexoses labeled in carbon atoms 1, 2, 5, and 6. Comparable data for mammals are insufficient to establish a glycolate pathway. Earlier work by Shemin, Weinhouse, and Sakami with labeled substrates established the rapid conversion of glycolate → glyoxylate → glycine ↔ serine (*70, 87, 88*). In mammals glycine and serine are gluconeogenic, and the serine is mainly metabolized via pyruvate, whereas in plants it is converted to glycerate.

III. Magnitude of the Glycolate Pathway

A rapid and large flow of carbon during photosynthesis through the glycolate pathway was long unexplained and ignored until the magnitude of photorespiration associated with this metabolism was

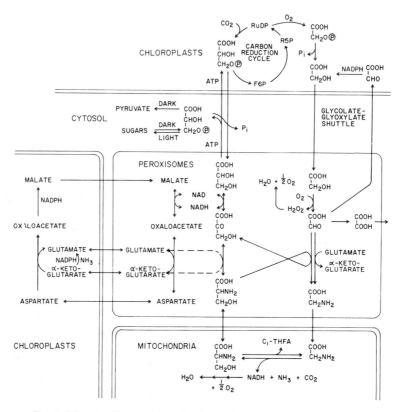

Fig. 1. The glycolate pathway in plants. RuDP, ribulose diphosphate.

found to reduce net rates of photosynthesis by as much as 40–50% (*30, 35, 80, 104*). In plants about 25% of the total ^{14}C fixed during a 2- to 10-minute exposure to $^{14}CO_2$ in the light is found in glycolate, glycerine, and serine. The pool sizes of P-glycolate, glycolate, and glyoxylate are small, and most of the carbon accumulates in the two amino acids glycine and serine. The rapidity and amount of ^{14}C-labeling of these two amino acids far exceed those of the other amino acids, even glutamate and aspartate in C_3-plants. Thus the biosynthesis of glycolate, glycine, and serine appears to represent a metabolic pathway which has other functions than the formation of these amino acids for protein synthesis. In $^{14}CO_2$ fixation experiments with end point measurements, it is not known how much more of the ^{14}C has also passed on through the glycolate pathway to sugars or has been oxidized back to CO_2. However, if the glycolate pathway were blocked by a sulfonate inhibitor of glycolate oxidase (*31, 100, 102*) or an isonicotinyl hydrazide inhibitor of glycine metabolism (*59, 60*), 50% or more of the fixed ^{14}C would then be found in glycolate. From many such experiments and from the magnitude of photorespiration, it appears that approximately half of the newly fixed carbon of photosynthesis may be metabolized by many C_3-plants via the glycolate pathway under optimum normal conditions.

In animals the presence of 5–100 μmoles of glycolate per 100 ml of blood ranks it as one of the organic acids found in significant amounts in blood, particularly of ruminants (*56, 98*).

IV. The Glycolate Pathway and Peroxisomes

Aerobic P-glycolate biosynthesis in chloroplasts initiates an aerobic metabolic sequence for glycolate metabolism (*80*) in the peroxisomes and mitochondria for the biosynthesis of glycine and serine, essential amino acids for protein, porphyrin and for C_1 biosynthesis (Fig. 1). Under normal environmental levels of light, CO_2, and O_2, excess production of P-glycolate seems to be unavoidable; as a result carbon also flows further from serine to glycerate in the peroxisomes and the glycerate in turn is used to regenerate sugars. In the absence or during insufficient photosynthetic P-glycolate biosynthesis, serine and glycine are formed from P-glycerate by an efficient anaerobic pathway. In the aerobic glycolate pathway energy is lost during P-glycolate formation, P-glycolate hydrolysis, glycolate oxidation, conversion of two glycines to one serine, in the reduction of hydroxypyruvate to glycerate and in rephosphorylation of glycerate. O_2 uptake occurs during P-glycolate biosynthesis, glycolate oxidation, and glycine oxidation. CO_2 is lost in the conversion of two glycines to serine, which may be repeated if the serine is reconverted to glycine and a C_1 compound (*39–41*). Thus the aerobic glycolate pathway of photorespiration is metabolically very wasteful compared to the anaerobic conversion of P-glycerate to serine. Only in leaves during photosynthesis does the massive flow of carbon through the glycolate pathway predominate, whereas serine formation from glycerate occurs in leaf peroxisomes at other times. These two peroxisomal routes to serine and glycine resides are initiated in the chloroplasts by the biosynthesis of P-glycolate and P-glycerate, which are then hydrolyzed by the specific phosphatases producing the free acids that are excreted and subsequently metabolized in the peroxisomes.

V. Environmental Factors Affecting Glycolate Biosynthesis

The four known major factors, which increase the relative amounts of glycolate formation by plants or algae, are low CO_2 concentration, high O_2 concentration, high light intensities (Fig. 2), and high pH of the media. A very extensive literature on these variables has been reviewed and will not be itemized (*26, 35, 79, 80,*

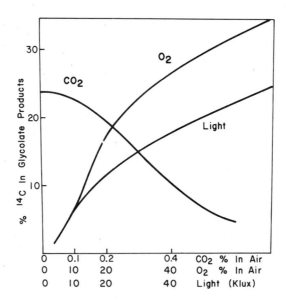

Fig. 2. Schematic representation of the effect of environmental factors upon glycolate formation by leaves. Each effect is considered with the other two held constant at that found in nature, namely, 21% O_2, full sunlight, and 0.03% CO_2 in air.

82, 91, 101, 104). The conditions now existing on this planet are quite favorable for photorespiration and unfavorable for maximum rates of photosynthesis.

A. Carbon Dioxide

Relative to the amount of CO_2 fixation, glycolate biosynthesis is maximal at low levels of CO_2 between 0.0 and 0.2% in air for algal cultures and higher plants. At the low CO_2 levels of the atmosphere (0.03%), glycolate products are among the major ones of photosynthesis. At higher CO_2 concentration, less glycolate is formed, photorespiration is less, net photosynthesis is greater, and sucrose is the dominant product. At very low or zero concentrations of CO_2, glycolate biosynthesis proceeds rapidly, drawing on the cells' reservoir of carbohydrate and converting it to CO_2. The carbohydrates newly formed during $^{14}CO_2$ fixation in a prior photosynthetic period are oxidized first back to $^{14}CO_2$ via glycolate. Evolution of CO_2 into CO_2-free air by plants in the light is measured by infrared absorption or as $^{14}CO_2$ and used to estimate the rate of photorespiration; this assay, however, does not take photosynthetic refixation of CO_2 into account and thus underestimates its magnitude. The same principle is the rationale for an assay of photorespiration based on the survival time of plants in CO_2-free air. C_3-plants which rapidly lose CO_2 from photorespiration can survive for only 5–6 days if the CO_2 in the atmosphere is continually being removed. This is not true for C_4-plants, which can refix internally nearly all of this CO_2. Likewise, algae in CO_2-free media biosynthesize and excrete glycolate in large amounts (*34*). All these experiments have been considered as indicative of a changeover from P-glycerate formation during CO_2 fixation to P-glycolate biosynthesis when there is inadequate CO_2. Since there is only 0.03% CO_2 in air and the effective concentration is even less in chloroplasts, and since there is 21% O_2 in air and nearer a 100% O_2 equivalent in the chloroplasts, the atmospheric situation favors glycolate formation. The stimulation of glycolate biosynthesis at low CO_2 can be explained as a physiological and perhaps enzymological competition between CO_2 and O_2 for RuDP carboxylase or RuDP oxygenase activity of fraction-1 protein of the chloroplast. The effect of low

CO_2 is also consistent with the hypotheses that excess reducing capacity will accumulate and may be dissipated by H_2O_2 formation from a chloroplast Mehler reaction and by glycolate oxidation. Proponents of the formation of glycolate by the oxidation of the TPP-C_2 complex of transketolase have shown that H_2O_2 can serve as the oxidant.

Most earlier experiments on $^{14}CO_2$ fixation by algae, higher plants, or isolated chloroplasts were done with low levels of CO_2. As the CO_2 levels were further reduced by its utilization, the product composition shifted toward glycolate, glycine, and serine formation. A full reevaluation of these experiments probably is not now necessary or feasible. Certainly some of the high yields of glycolate production by algae can be explained by experimental conditions of very low CO_2 and high O_2.

B. Oxygen

Glycolate formation and photorespiration are enhanced with increasing O_2 concentration up to 100%, whereas dark mitochondrial respiration is saturated above 2% O_2. O_2 uptake during photorespiration occurs at three sites, P-glycolate synthesis, glycolate oxidation, and glycine oxidation (Fig. 1). P-glycolate biosynthesis in the chloroplast by RuDP oxygenase has a K_m (O_2) of about 7.5×10^{-4} M, which is obtained by aerating media with 66% O_2 in air. Glycolate oxidation in the peroxisomes by the flavoprotein, glycolate oxidase, is effectively saturated by about 60% O_2 (45). Mitochondrial glycine oxidation is linked with NADH oxidation and should be saturated by low levels of O_2, as was observed with isolated mitochondria (10). However, high oxygen concentration also stimulates glycine decarboxylation in leaves, for unknown reasons (42). The exact O_2 concentration in solution in different parts of a leaf is not known, but in the light there should be a gradient from the highest amount in the chloroplast down to that obtained by equilibrium with 21% O_2 in the air at the stomatal cavity. In the dark only 21% oxygen in air is present. Thus the high O_2 concentration generated by photosynthesis increases the rate of photorespiration and exerts an inhibitory effect on net photosynthesis. The most important effect of high O_2 concentration seems to be on the biosynthesis of glycolate, although high concentrations of oxygen are also essential for glycolate metabolism. Physiologically the oxygen effect of glycolate biosynthesis and photorespiration has been confirmed by a 50% increase in the growth rate of C_3-plants in 2–5% O_2 (11) or the complete inhibition of growth of plants in oxygen of over 30%. This phenomenon was first described by Warburg in 1920 as an O_2 effect on the inhibition of photosynthesis (26, 27). All these experiments indicate that normal growth of plants in air is severely limited by 21% oxygen because of glycolate biosynthesis. There is also a correlation between O_2 solubility, temperature optima for growth of plants, and the magnitude of photorespiration. O_2 solubility from air into water decreases from about 310 μmoles per liter at about 15°C to 210 μmoles per liter at 35°C. In general, C_3-plants with high rates of photorespiration have temperature optima from 15°C to 25°C where more O_2 from photosynthesis would remain in solution. C_4-plants with less photorespiration have temperature optima from 25° to 35°, where O_2 retention may be as much as 30% less.

C. pH

Glycolate biosynthesis *in vivo* is favored by a high pH. With *Chlorella* in media above pH 8.0, glycolate is the major product of photosynthesis (52), and glycolate formation at pH values below 6 to 7 becomes nil (84). Glycolate formation by broken cell fractions containing chloroplasts is also greatly favored at pH values above 8.5 (23). These results can be explained by the high pH optimum of 9.3 for the RuDP oxygenase and the decrease in RuDP carboxylation by the near absence of CO_2 in this pH range. Thus CO_2 fixation decreases rapidly above pH 8 due to the decrease in CO_2 and the unfavorable pH for the carboxylase, whereas P-glycolate biosynthesis increases.

D. Light

The rate increases of glycolate formation with increasing light intensity and does not become saturated even with full sunlight. This is consistent with photorespiration as a process, in part, to utilize excess photoassimilatory power which increases with light intensity. Conversely the steady state level of glycolate in *Chlorella* decreases very rapidly in the dark, similarly to the pool of RuDP (*55*). Light dependency is explained in the regeneration of the substrates, such as RuDP, for P-glycolate or glycolate biosynthesis, in the same manner that CO_2 fixation depends ultimately on light generated ATP and $NADPH_2$. Light dependency for glycolate biosynthesis is consistent with the absence of photorespiration phenomena in the dark. If glycolate were formed from the peroxidation of $TPP-C_2$, the need for high light intensity could be explained by the production of excess reducing capacity, which in isolated chloroplasts is autooxidized by oxygen to form H_2O_2 (the Mehler reaction (*26*). RuDP carboxylation to P-glycerate or oxygenation to P-glycolate are dark reactions. The carboxylation reaction during photosynthesis is saturated at intermediate light intensities. The continued increase in rate of P-glycolate biosynthesis with increasing high light intensity may be related to the high pH optimum of the RuDP oxygenase reaction. The chloroplast stroma become increasingly alkaline with increasing light intensity as protons are transported through the membrane of the thylakoids into their inner space (*36*, *49*). About 0.2 mole H^+ per mole chlorophyll are pumped in and give rise to a pH difference across the thylakoid membrane of 3 pH units (*69*). If rates of carboxylation versus oxygenation of RuDP were measured between pH 7.8 and 9.0, one should pass into a range above pH 8.3, where carboxylation becomes very difficult and oxygenation will accelerate (Fig. 3). A light-driven H^+ gradient should produce a higher pH in the stroma, and as a result less CO_2 would be available for carboxylation and the pH would favor the oxygenase reaction. The inward translocation of protons has been reported to be accompanied by an outward movement of potassium and magnesium ions (*21*, *54*) and inward movement of chloride ions (*19*). On the other hand, the formation of 2 moles of acid during RuDP oxidation in the stroma and a further mole of acid during P-glycolate hydrolysis would also neutralize the alkalinity from the inward movement of protons and shift the reaction back toward its initial state.

Glycolate biosynthesis is curtailed by inhibitors of electron transport such as CMU (*27*, *73*) or Mn^{2+} deficiency (*33*, *77*). The

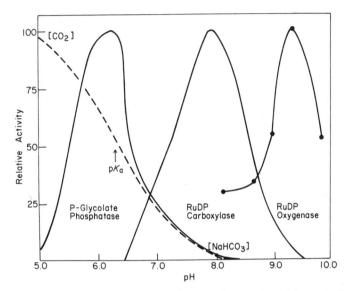

Fɪɢ. 3. Effect of pH upon CO_2 concentration and activity of ribulose diphosphate (RuDP) carboxylase, RuDP oxygenase, and phosphoglycolate phosphatase.

specificity of inhibition of glycolate biosynthesis by Mn^{2+} deficiency as compared to CO_2 fixation has not been explained. The phenomenon might be considered as equivalent to lowering the light intensity or raising the effective CO_2 concentration which preferentially reduce glycolate biosynthesis.

VI. Ribulose Diphosphate Oxygenase-Catalyzed Formation of Phosphoglycolate

A. Products of the Reaction

Homogeneous preparations of fraction-1 protein from either soybean and spinach leaves catalyze both a RuDP carboxylase and a RuDP oxygenase reaction (*3, 12, 13, 46, 50*). In the carboxylase reaction, CO_2 is added at carbon 2 of the RuDP followed by hydrolytic cleavage between carbons 2 and 3 to form two molecules of 3-P-glycerate. The same enzyme can catalyze also the addition of O_2 at carbon 2 of the RuDP to form the hypothetical superoxide intermediate of RuDP, shown in Eq. (1), which is then hydrolyzed.

$$(1)$$

In the oxygenase reaction the products are P-glycolate from carbons 1 and 2 and 3-P-glycerate from carbons 3, 4, and 5 of the RuDP. The oxygenase has been assayed manometrically, and characterized about as well as the carboxylase reaction. The identity of the products was best proved by the use of $^{18}O_2$ and combined gas chromatography and mass spectrometry of the trimethylsilyl derivatives (46). One atom of molecular $^{18}O_2$ was incorporated into the carboxyl group of the P-glycolate, none into the C_2-alcoholic group of P-glycolate, and none into 3-P-glycerate. When the reaction was run with $^{18}O_2$-labeled water, shown by an asterisk in Eq. (1), the carboxyl group of 3-P-glycerate was labeled as predicted by hydroxyl ion attack on carbon 3 of the RuDP intermediate. Andrews et al. (3), using RuDP-^{14}C, also found only these two products of the enzymatic reaction by paper chromatography. Ogren and Bowes (12, 13, 50) first predicted the formation of P-glycolate by measuring the formation of glyoxylate as its phenylhydrazone, after treatment of the reaction products with P-glycolate phosphatase and glycolate oxidase. That a similar reaction occurs in vivo was proved by exposing a spinach leaf to the $^{18}O_2$ tracer and identifying the tagged products by mass spectrometry (2). In vivo also one atom of O_2 was incorporated only into the carboxyl group of glycine and serine derived from the P-glycolate; none was incorporated into P-glycerate. *In vivo* the glycine and serine pools are labeled to a maximum specific activity with either $^{14}CO_2$ or $^{18}O_2$ within 2–5 minutes. Calculations by Andrews *et al.* (*2*) did not indicate a full one atom of $^{18}O_2$ incorporation into glycine and serine *in vivo*, because of inadequate correction for the natural abundance of the silicon isotopes of the silylated derivatives, but this was corrected in a subsequent paper (*46*). Thus the rapid $^{18}O_2$ labeling of the carboxyl group of P-glycolate and its metabolic products seems to be a unique characteristic of glycolate biosynthesis during photorespiration in plants. Other theories for glycolate formation must also account for the labeling patterns with both $^{18}O_2$ and $^{14}CO_2$.

P-Glycolate formation from the top two carbon atoms of RuDP is consistent with rapid uniform labeling of both during $^{14}CO_2$ fixation. The 3-P-glycerate from the lower 3 carbons would retain carboxyl labeling as carbon 3 of the RuDP is initially most heavily labled. This 3-P-glycerate would be used to regenerate more RuDP, as shown in Fig. 4. The established photosynthetic carbon cycle fixes 6 CO_2, forms 1 hexose, and regenerates 6 RuDP (*5*). Alternatively, in low CO_2 near the compensation point when net photosynthesis is zero and no hexose-P is removed, the cycle can be written to form no net hexose but to regenerate 9 RuDP instead. Three of the RuDP undergo oxidation to three P-glycolates, as well as the extra three 3-P-glycerate that are needed for the RuDP regeneration. Further modification (not shown in Fig. 4) occurs in the absence of enough CO_2 to produce sufficient 3-P-glycerate for the regeneration of the RuDP during continued photorespiration. This necessitates drawing upon the hexose reserves of the chloroplast or cell for P-glycerate formation. Rapid glycolate biosynthesis and CO_2 loss during photorespiration in the absence of CO_2 soon kills the plant because of this continuing RuDP oxygenase activity in the light.

B. Enzyme Properties

The RuDP oxygenase and carboxylase activities copurified with fraction-1 protein from soybean (*13*) or spinach leaves (*3*) through procedures involving $(NH_4)_2SO_4$ fractionation, sucrose density gradient centrifugation, and DEAE-cellulose chromatography. All efforts to separate the two activities were unsuccessful, and thus it appears that the same enzyme catalyzes both reactions, which

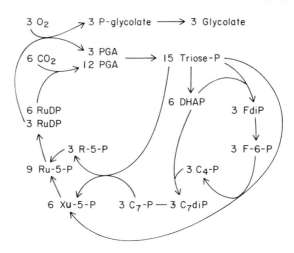

FIG. 4. The photosynthetic carbon cycle near the CO_2 compensation point. P-glycolate, phosphoglycolate; PGA, phosphoglyceric acid; RuDP, ribulose diphosphate; R-5-P, ribose 5-phosphate; Xu-5-P, xylulose 5-phosphate; F-6-P, fructose 6-phosphate; FdiP, fructose-1,6-diphosphate; DHAP, dihydroxyacetone phosphate.

are indeed quite similar. The ratio of the two activities, carboxylase/oxygenase, each measured at optimum conditions was 0.25 in crude extracts but increased to 0.50 or higher during purification and storage of the protein. This may indicate a greater stability of the oxygenase activity than the carboxylase activity. The significance of these differences in stability is unclear, but may indicate that the oxygenase is only a partial reaction relative to the carboxylase. The K_m (RuDP) for both the carboxylase and oxygenase is about the same, being 1 to 2×10^{-4} M. The K_m (CO_2) at 4.5×10^{-4} M (18) is also similar to the K_m (O_2) at about 7.5×10^{-4} M (2). In order to obtain this molarity of O_2 in water, the solution has to be aerated with 66% O_2 in air. The oxygenase was not saturated by aeration with 100% O_2, and when aerated with air the rate was 37% of that in pure oxygen. Thus neither the carboxylase, even when activated (see Chapter by Buchanan and Schürmann), nor the oxygenase activity is saturated by normal *in vivo* concentrations of these gases. Consequently great differences in rate of the two reactions occur in nature as the atmosphere is varied.

The pH optimum of the oxygenase is high, 9.3–9.5, with activity tailing off below pH 8 (Fig. 3) (3). Values below 8 were not quantitated because CO_2 concentration in the media increased and the carboxylase activity accelerated to use up the common substrate. The pH optimum of the carboxylase is at 7.8, and it is nearly inactive at pH 8.5. These two different pH optima of the purified enzyme may not be entirely representative of *in situ* conditions, but the results are consistent with a high pH optimum for glycolate biosynthesis. The RuDP carboxylase reaction is specific for CO_2, not for bicarbonate (18). In Fig. 3 is also shown the dissociation curve for the reaction of CO_2 and base to form bicarbonate in order to emphasize that the CO_2 concentration at pH 8.0 and above is nil, so that the carboxylase reaction above pH 7.8 *in vivo* is very limited. Although carbonic anhydrase of the chloroplast can accelerate the rate of reaching the equilibrium between CO_2 and bicarbonate, it cannot alter the pK_a of this equilibrium.

Further studies will be necessary to elucidate the mechanism of the RuDP oxygenase reaction. The reaction seems to be similar to the addition of CO_2 to RuDP during P-glycerate formation, and it is inhibited by CN^- (3) as is the carboxylase (95). Lorimer *et al.* (46) proposed the superoxide intermediate of RuDP, as shown in Eq. (1), because $^{18}O_2$ was incorporated only at carbon 2 of the RuDP. The enzyme has been reported to contain one atom of copper (96), whose removal does not affect the carboxylase activity. The preparations used by Lorimer *et al.* (46) did not contain copper, so that its logical involvement in the oxygenase reaction could not be suggested.

C. Detection of Phosphoglycolate *in Vivo*

P-glycolate is rapidly labeled during $^{14}CO_2$ fixation by algae and higher plants (6, 9, 103) and by isolated chloroplasts (38). In kinetic studies of the steady-state levels of intermediates of CO_2 fixation, Bassham's group found a rapid decrease in RuDP and an increase in P-glycolate simultaneously or preceding the rise in pool size of glycolate, when the system was disturbed by increasing the O_2 concentration to 100% (Bassham, personal communication). It is essential in these investigations to kill the tissue by procedures that inactivate the heat-stable P-glycolate phosphatase and to use chromatographic procedures that adequately separate P-glycolate and P-glycerate. P-glycolate has not been reported in tissues other than plants.

VII. Phosphoglycolate Phosphatase and Phosphoglycerate Phosphatase

Chloroplasts rapidly synthesize P-glycolate and P-glycerate. Whereas P-glycerate accumulates in the cell as an essential component of the photosynthetic carbon cycle and the Embden-Meyerhof pathway, P-glycolate is a potent inhibitor (K_i of 2.2×10^{-6} M) of triose-P isomerase (97) and to a lesser extent of RuDP carboxylase (H. S. Ku and N. E. Tolbert, unpublished). Both phosphate esters can be hydrolyzed by different specific chloroplast enzymes, P-glycolate phosphatase (1, 67) and P-glycerate phosphophatase (63, 64). Free glycolate and glycerate are excreted from the chloroplasts, and further metabolized in the peroxisomes (80) (Fig. 1). Amounts of both phosphatase activities vary among plants, but range from 1 to 20 μmoles min^{-1} mg^{-1} chlorophyll (64). The activities seem adequate to account for the rate of formation of glycine and serine during photosynthesis; in particular P-glycolate phosphatase is present in excess. In general C$_3$-plants with high rates of photorespiration contain approximately 4 times as much P-glycolate phosphatase as P-glycerate phosphatase (64). In contrast, in the mesophyll cells of C$_4$-plants with less photorespiration, this ratio of 4 is reversed, because there is less P-glycolate phosphatase. However, in the bundle sheath cells of these same C$_4$-plants P-glycolate phosphatase is more active than P-glycerate phosphatase.

Regulation of these two different phosphatases *in vivo* would seem essential, but only their isolation and some of their properties have been described (Table I). That the hydrolysis of two nearly identical phosphate esters from the initial reaction products of photosynthesis occurs by different enzymes indicates an ability to control each separately, i.e., hydrolyze P-glycolate as fast and completely as possible, at the same time permitting only a limited hydrolysis of P-glycerate. Because of the relatively large pool of 3-P-glycerate in photosynthetic tissue, the 3-P-glycerate phosphatase must be under strict control, yet it has never been understood why many

TABLE I
PROPERTIES OF PHOSPHOGLYCOLATE PHOSPHATASE AND
PHOSPHOGLYCERATE PHOSPHATASE

Property	P-glycolate phosphatase	3-P-glycerate phosphatase
Substrate specificity	P-glycolate	3-P-glycerate, P-enol pyruvate
Cellular location	Chloroplasts	Chloroplast and cytosol
pH optima	6.3 with Mg^{2+} 5.3 with Zn^{2+}	5.9–6.3
Molecular weight	$\sim$60,000[a]	$\sim$160,000
K_m	$2.5 \times 10^{-3}\ M$	$8 \times 10^{-4}\ M$
Effect of dialysis or EDTA	Inactivates[a]	No effect
Cation requirement	Mg^{2+} or Zn^{2+}	None
Stabilizer	Tricarboxylic acid[a]	None
Inhibitors	None	Zn^{2+} or Cu^{2+}
Lability at 55°C	Stable[a]	Inactive

[a] Unpublished.

plants accumulate immense amounts of free glycerate (carboxyl labeled similar to 3-P-glycerate) during photosynthesis, but on a time scale (minutes) considerably slower than the $^{14}CO_2$ saturation of the 3-P-glycerate pool (31, 47). Rapid formation of labeled P-glycolate during $^{14}CO_2$ fixation has been observed by classical chromatographic procedures (9), but usually the P-glycolate pool appears much smaller than the amount of P-glycerate. P-glycolate phosphatase is quite stable to heat (1). In most experiments using boiling alcohol or water to kill the plant tissue this phosphatase is not inactivated (85), and as a result the amount of P-glycolate present may be severely underestimated. So far there is no evidence for the regulation of the large amount of P-glycolate phosphatase, and the absence of a large reservoir of P-glycolate is consistent with its rapid removal in order to prevent inhibition of carbohydrate metabolism.

Common properties of the two specific phosphatases are the similarity of the chemical structure of the substrates, similar pH optima of 6.3, which is unusually neutral for phosphatases, similar location in the chloroplast, and similar development during greening of leaves. The two phosphatases differ markedly in their cation requirement and stability, permitting assay of one in the presence of the other. P-glycolate phosphatase requires Mg^{2+} or Zn^{2+}, is inactivated by EDTA or dialysis, and is cold labile and heat stable. P-glycerate phosphatase has no demonstrable requirement for a cation, is stable to EDTA and dialysis, and is in fact inactivated by Zn^{2+} or heat. P-glycolate phosphatase, as isolated from leaf homogenates, has many moles of ionically bound tricarboxylic acid per mole (citrate from tobacco leaves or *cis*-aconitate from wheat leaves) (*1*). The acetone powder of the purified phosphatase must be dried slowly near room temperature but is then stable for years. Passage of the enzyme through a DEAE-cellulose column partially removes the tricarboxylic acids, and then the protein, though initially active, is very labile, and stability can be restored specifically by the tricarboxylic acids. *Cis*-Aconitate is a competitive inhibitor, K_i (2×10^{-3} M).

VIII. Glycolate from the Oxidation of a Thiamine Pyrophosphate–C_2 Complex

Much literature on the oxidation of the transketolase C_2-addition complex of dihydoxyethylthiamine pyrophosphate (TPP-C_2) *in vitro*, shown in Eq. (2), has accumulated in the past 20 years and has been discussed in reviews (*4, 14, 26, 90, 93*).

$$\text{fructose-6-P} \xrightarrow{\text{transketolase}} \overset{\overset{\text{TPP}}{|}}{\underset{\underset{\text{CH}_2\text{OH}}{|}}{\text{H—C—OH}}} \xrightarrow[\substack{\text{Fe}^{3+} \\ \text{quinone}}]{\text{H}_2\text{O}_2} \overset{\text{COOH}}{\underset{\text{CH}_2\text{OH}}{|}} + \text{TPP} \qquad (2)$$

TPP-C_2, as an intermediate in the transketolase reaction, is oxidized nonenzymatically to glycolic acid by a variety of oxidants such as H_2O_2, Fe^{3+}, and dichlorophenolindophenol. This has been an attractive hypothesis for the mechanism of glycolate biosynthesis, since the C_2 moiety would be uniformly labeled with ^{14}C from known intermediates of the photosynthetic carbon cycle. Recently, extensive studies by Gibbs and associates (*26, 27, 57, 75*) have proved that isolated chloroplasts, to which transketolase and fructose-6-P had been added can oxidize the intermediate to glycolate in the light at rates approaching that required for photorespiration.

The oxidant is generated by the photosystems and is presumably H_2O_2, in part, since added catalase greatly limited glycolate production. H_2O_2 in turn could be generated from the oxidation by molecular oxygen of reduced ferredoxin (78), the flavoprotein ferridoxin-TPN reductase, or the primary electron acceptors from either photosystem I or II. These reactions would be favored by high O_2 concentrations and might involve a superoxide intermediate. Several inhibitors of electron transport preferentially reduced glycolate formation by isolated chloroplasts, as compared to CO_2 fixation, as would be the case if H_2O_2 formation was the result of excess reducing capacity. Further evaluation of this hypothesis has been discussed in the cited references by Gibbs' group, and there is no doubt that they have shown the feasibility of the mechanism with isolated chloroplasts. However, definitive experiments *in vivo* to establish this mechanism have not been devised. Several of these experimental parameters may not occur *in situ*, particularly the formation of H_2O_2 in chloroplasts. The hypothesis does not account for $^{18}O_2$ incorporation into the carboxyl group of glycolate, as the oxidation is expected to form 2 moles of H_2O from the H_2O_2. Further, this mechanism does not account for P-glycolate formation and the presence of the very active P-glycolate phosphatase of the chloroplasts. Experiments on the oxidation of the TPP-C_2 complex by the chloroplasts were carried out at the same or lower pH than that for the RuDP carboxylation reaction, rather than at higher pH as observed *in vivo* for glycolate biosynthesis. The preferential inhibition of glycolate formation by adding certain compounds to the isolated chloroplasts is relevant to the *in vitro* system but does not prove that these inhibitors function *in vivo* in this way.

Whittingham *et al.* (90) observed that glucose-2-^{14}C formed carboxyl-labeled glycolate and that glucose-1-^{14}C formed glycolate-2-^{14}C when fed to *Chlorella* in the light and proposed that these results were evidence for glycolate formation from the oxidation of a TPP-C_2 arising from fructose-6-P. The data may equally well be explained as evidence for transketolase-catalyzed formation of Xu-6-P and conversion of that to RuDP, followed by oxidation to the specific labeled P-glycolate.

In a series of papers, Holzer's group has shown that the TPP-C_2 from hydroxypyruvate decarboxylation by pyruvate oxidase is oxidized enzymatically to glycolate by ferricyanide or dichlorophenol-indophenol (*24*). There is also no reason to support this route *in vivo* for glycolate biosynthesis during photosynthesis, particularly since the glycerate dehydrogenase for hydroxypyruvate formation is located in the peroxisomes, yet glycolate synthesis occurs in the chloroplasts. However, TPP-C_2 oxidation is a working hypothesis for glycolate biosynthesis, and it is of particular interest for non-photosynthetic tissues without RuDP oxygenase.

IX. Other Possible Mechanisms

A. Glycolate from Glyoxylate

Glycolate formation from glyoxylate by the activity of a reductase is a possibility in some instances, but this route is not considered to be a major pathway for glycolate biosynthesis during photosynthesis. In this hypothesis the enzyme investigated most is a reductase located in leaf peroxisomes (*80*). It was first described as a glycerate dehydrogenase (*76*) and as a glyoxylate reductase (*99*). That it does not function as a glyoxylate reductase but as a hydroxypyruvate reductase or glycerate dehydrogenase (Fig. 1) was concluded by Tolbert *et al.* (*83*) for several reasons. (a) The K_m (glyoxylate) is 2×10^{-2} M; (b) there is no evidence for the reduction of glyoxylate-^{14}C to glycolate by plants *in vivo;* and (c) peroxisomal or photorespiration does not occur in the dark as would happen if this enzyme were linked to glycolate oxidase to function as a terminal oxidase system.

Chloroplasts contain a specific glyoxylate reductase which utilizes $NADPH_2$ and has a K_m (glyoxylate) of 10^{-4} M (*83, 105*). However, the total activity of this enzyme in chloroplasts is so low (unpublished) that it could not account for the rates of glycolate biosynthesis. Formation of glycolate from glyoxylate in chloroplasts establishes a potential shuttle between the chloroplasts and peroxisomes that could remove excess reducing capacity through glycolate oxidation in the peroxisomes (Fig. 1) (*80*). This cyclic process would not result in net glycolate synthesis or any CO_2 formation, and net O_2 exchange between glycolate oxidation and photolysis of water would be zero. Thus it is logical that the chloroplast NADPH glyoxylate reductase is of low controllable activity and probably closely regulated. *In vivo* experiments to prove existence of this terminal oxidase system between the two organelles, chloroplasts and peroxisomes, is lacking.

Sufficient glyoxylate formation to account for the glycolate flux by other mechanisms seems unlikely, particularly since it would more likely be converted to glycine. Glyoxylate is an intermediate in tissues with the glyoxylate cycle, but this does not include green leaves or animal tissues, and furthermore, glyoxylate from the glyoxylate cycle is all reutilized for malate synthesis, while there is little evidence of its net reduction to glycolate. In fact glyoxysomes of seedlings contain a large amount of the glutamate-glyoxylate aminotransferase (*73*), which suggests its conversion to glycine instead. Some glyoxylate is formed as a degradation product of certain metabolites, such as allantoate and hydroxyproline. The formation of glyoxylate and glycolate from oxalate has been reported for certain bacteria (*61*), but not for plant or animal tissue. Glyoxylate formation cannot occur from glycine, since the two glutamate–glyoxylate and serine–glyoxylate aminotransferase reactions are "physiologically" irreversible (*66*). Rather glycine is metabolized in the mitochondria by oxidation to CO_2 and a C_1 moiety concurrently with NAD reduction and subsequent electron transport leading to ATP synthesis and O_2 uptake.

Lactate dehydrogenase in animal tissue (there is little or none in leaves) catalyzes the rapid reduction of glyoxylate to glycolate (*68, 71*). This action accounts for earlier feeding data showng glycolate-^{14}C formation from glyoxylate-^{14}C by perfused rat liver (*87*). An interpretation of these results remain unclear, since it is not known whether significant amounts of glyoxylate are formed in liver. However, liver and kidney have the potential to dispose of excess reducing capacity by a therminal oxidative system between the lactate dehydrogenase of the cytosol and the α-hydroxy acid oxidase of the peroxisomes. The peroxisomal oxidase and catalase system can oxidize the α-hydroxy acids, lactate and glycolate, to pyruvate and glyoxylate, which in turn would be reduced by the NADH-linked lactate dehydrogenase. Also, erythrocytes contain a glyoxylate reductase (*94*), as well as the glycolate oxidase (unpublished) and catalase. Though this potential terminal oxidase system exists in mammals, because of the presence of the substrates and enzymes, there is no physiological evidence for the hypothesis nor clue as to the source of the two carbon compounds.

B. Glycolate from a CO_2 Plus CO_2 Condensation

The formation of glycolate from a CO_2–CO_2 condensation has been proposed (77, 103). Zelitch (103, 104) claimed that the initial specific activity of glycolate formed in tobacco leaves during photosynthesis was greater in the presence of $^{14}CO_2$ than that in the carboxyl group of 3-P-glycerate. However, this was not confirmed by Tolbert's group (31), with or without sulfonate inhibitors to promote the accumulation of glycolate. They found that the carboxyl group of P-glycerate had higher specific activity than the glycolate during the first 60 seconds. For algae the kinetics of ^{14}C labeling of glycolate with time have clearly shown that it is formed after the labeling of 3-P-glycerate (16, 52). In these experiments glycolate reached maximum specific activity in less than 1 minutes, emphasizing that it is biosynthesized rapidly yet does not accumulate, for it is converted to glycine and serine. In short-time, 4-second, experiments with either *Chlamydomonas* or tobacco leaves, glycolate was initially labeled more in C-2 than in the carboxyl group (15, 32). After 12 seconds the glycolate was uniformly labeled. These results are thought to be consistent with the predicted initial labeling pattern of C_1 and C_2 of RuDP, but they are certainly inconsistent with a direct fixation of CO_2 to form glycolate. Other factors in glycolate biosynthesis not explained by the direct CO_2 fixation hypothesis are incorporation of $^{18}O_2$ only into the carboxyl group and the absence of any enzymatic mechanism to catalyze such a feat.

C. Glycolate from Acetate

Although occasionally proposed from *in vivo* metabolism studies, enzymatic conversion of acetate to glycolate or vice versa has not been demonstrated in a convincing manner, and thermodynamically it is not feasible by pyridine nucleotide-linked oxidative reductases. Several publications on apparent initial rapid conversion of labeled acetate into glycolate have been reported for algae, but the possibility that the labeled glycolate was an impurity in the acetate was not excluded (28). Papers on long-term acetate-^{14}C feeding followed by detection of glycolate do not establish a mechanism of formation for metabolites with a rapid turnover. In microorganisms assimilation of acetate by the glyoxylate cycle is a possible mechanism for glycolate biosynthesis from the glyoxylate formed from the isocitrate lyase reaction. Though this hypothesis may account for ^{14}C labeling of a glycolate pool when feeding acetate-^{14}C, there is no evidence for net synthesis of glycolate by the glyoxylate cycle.

X. Glycolate Excretion after Formation

After glycolate biosynthesis in the chloroplast, it is excreted into the cytosol. Isolated plant chloroplasts excrete glycolate, while retaining certain of the phosphate esters of the photosynthetic carbon cycle (38). Laboratory cultures of unicellular green algae, such as *Chlorella* and *Chlamydomonas,* excrete several percent (2–12%) of their total photosynthate into the medium as glycolate (43, 84, 92). These observations have been extended to most unicellular algae in nature, where many factors affect the exact magnitude of glycolate excretion, and its significance is unclear (82). Both the higher plant chloroplast and algae contain large amounts of P-glycolate phosphatase which may be involved in this excretatory process. Glycolate excretion by algae is dependent upon the presence of bicarbonate ions and oxygen, suggesting an active transport phenomenon (84). Algae do not excrete all the glycolate formed, but convert part of it to glycine and serine by a pathway similar to that shown in Fig. 1. The major difference between the algal glycolate pathway and that of the higher plant is that in the algae the glycolate is oxidized by a dehydrogenase which is not linked to O_2 (48). As a result, most unicellular green algae do not form H_2O_2 by this pathway, have relatively low catalase content, and may not contain peroxisomes (82).

XI. Speculation and Discussion

Although the title of this paper is "Glycolate Biosynthesis," in actuality the important step is P-glycolate biosynthesis followed by its enzymatic hydrolysis to glycolate. Retention of the term "glycolate biosynthesis" seems justified, as is the case for reviews on glucose biosynthesis, where the last step is hydrolysis by a phosphatase.

Our concept of glycolate biosynthesis is that it is formed by a RuDP oxygenase activity of protein fraction 1 of leaves, which is also an RuDP carboxylase. This enzyme is specific for RuDP but lacks specificity for the other substrate, adding CO_2 or O_2 to carbon atom 2 of the RuDP. Addition of CO_2 leads to net photosynthesis; addition of O_2 gives rise to P-glycolate biosynthesis and photorespiration. Thus photosynthesis and photorespiration represent a competition between CO_2 and O_2 of the atmosphere for this single key reaction for plant growth. Concentrations of both gases in solution from an equilibrium with air are insufficient and are below K_m concentrations observed for the isolated enzyme. The present balance of CO_2 and O_2 in air is a manifestation of this competition.

The reason for the biosynthesis of P-glycolate by fraction-1 protein should be considered further. Data may indicate that this enzyme acts similarly in all aerobic photosynthetic tissues, but so far the reaction has been studied only as isolated from spinach (3) and soybean (13) leaves. P-glycolate biosynthesis and P-glycerate formation during photosynthesis by the same enzyme may be intimately related with the mechanism of the reaction of CO_2 fixation. This new insight should provide further clues as to the mechanism of action of CO_2 fixation by the carboxylase, whose K_m (CO_2) of 4.5×10^{-4} M and pH optimum at 7.8–8.0 (where the CO_2 is mostly all bicarbonate) are unfavorable for photosynthesis. It seems possible that simultaneous P-glycolate formation with CO_2 fixation may be an essential feature of photosynthesis, perhaps to concentrate CO_2 at the stroma chloroplast location of the enzyme. In C_4 plants with little CO_2 loss during photorespiration, the amount of glycolate synthesis and glycolate oxidase appear to be less (65), particularly in sorghum (53). In these plants the C_4-dicarboxylic acid cycle functions for CO_2 storage. Further comparisons between glycolate biosynthesis and photorespiration in C_3-plants with CO_2 fixation in C_4-plants seem necessary.

Several ideas and properties of P-glycolate phosphatase suggest that this enzyme and glycolate biosynthesis may be involved in membrane transport phenomena of the chloroplasts. The involvement of other phosphate esters and phosphatases in membrane transport has set such a precedent. It is very readily solubilized from chloroplasts, like RuDP carboxylase, but if intact chloroplasts are isolated in media containing sorbitol or mannitol the phosphatase remains in part with the plastids. The enzyme is in all green algae which we have tested, and it can be removed from unicellular green algae by osmotic shock that does not break the cell.

In the biosynthesis of glycolate, RuDP oxygenase for P-glycolate formation functions optimally at pH 9.3 to 9.5, while the phosphatase has at pH maximum of 6.3 (with Mg^{2+}) to 5.3 (with Zn^{2+}) (Fig. 3). Thus a pH gradient of over 3 units is demanded by these two consecutive reactions. Their location on two different sides of a membrane or area of a chloroplast seems necessary to accommodate this pH difference. We speculate that if glycolate biosynthesis is related to membrane transport, bicarbonate transport is most likely to be involved. Some reasons for this hypothesis are that glycolate biosynthesis is directly related to CO_2 availability in photosynthesis, that glycolate biosynthesis creates acid equivalents, and that algae do not excrete glycolate in the absence of CO_2 or bicarbonate.

Plant physiologists are investigating the effect of the quality of light on glycolate biosynthesis, metabolism, and respiration. The results cannot be explained by current knowledge about glycolate biosynthesis. Green algae, *Chlorella* and *Anacystis*, form and excrete glycolate in red or white light, but in blue light they accumulate little and excrete none in short termed experiments of minutes (7, 22, 44). Spruce leaves have a higher rate of photorespiration in blue light (58), and higher plants form more amino acids, particularly glycine and serine (literature not summarized). After long-term adaptation over several days to blue light, *Chlorella* accumulate a much larger reservoir of glycolate-^{14}C during $^{14}CO_2$ fixation (33). Blue light stimulation of respiration has been attributed to stimulation of the respiratory flavoproteins (72).

Although glycolate and glyoxylate are reported to be present in blood and urine, information on how they may be formed in mammals is very limited, and no large, specific route is known as in plants. Some glycolate may be a dietary constituent from leaves, but this is likely to be a minor source since glycolate does not accumulate, owing to its rapid conversion to glycine. In this review isolated facts that relate to the glycolate pathway in blood erythrocytes were also mentioned. These included the presence of significant amounts of glycolate in blood, and catalase, glycolate oxidase, and glyoxylate reductase in erythrocytes. The report of RuDP carboxylase (25) and of P-glycolate (51) in blood so far cannot be confirmed (unpublished). The α-hydroxy acid oxidase in peroxisomes of liver and kidney is similar to that in leaf peroxisomes. The function of mammalian peroxisomes remains a major enigma in biochemistry (80). Further comparisons between the aerobic glycolate pathway of metabolism in leaves with that in mammalian tissues are to be expected.

References

1. Anderson, D., Ph.D. Thesis, Michigan State University, East Lansing (1969).
2. Andrews, T. J., Lorimer, G. H., and Tolbert, N. E. Biochemistry 10, 4777–4782 (1971).
3. Andrews, T. J., Lorimer, G. H., and Tolbert, N. E., Biochemistry 12, 11–18 (1973).
4. Bassham, J. A., Annu. Rev. Plant Physiol. 15, 101–120 (1964).
5. Bassham, J. A., Benson, A. A., Kay, L. D., Harris, A. Z., Wilson, A. T., and Calvin, M., J. Amer. Chem. Soc. 76, 1760–1770 (1954).
6. Bassham, J. A., and Kirk, M., Biochem. Biophys. Res. Commun. 9, 376–380 (1962).
7. Becker, J. D., Döhler, G., and Egle, K., Z. Pflanzenphysiol. 58, 212–221 (1968).
8. Benson, A. A., and Calvin, M., J. Exp. Bot. 1, 63–68 (1950).
9. Benson, A. A., Bassham, J. A., Calvin, M., Hall, A. G., Hirsch, H. E., Kawaguchi, S., Lynch, V., and Tolbert, N. E., J. Biol. Chem. 196, 703–716 (1952).
10. Bird, I. F., Cornelius, M. J., Keys, A. J., and Whittingham, C. P., Phytochemistry 11, 1587–1594 (1972).
11. Björkman, O., Hiesey, W. M., Nobs, M., Nicholson, F., and Hart, R. W., Carnegie Inst. Wash., Yearb. 66, 228–232 (1968).
12. Bowes, G., and Ogren, W. L., J. Biol. Chem. 247, 2171–2176 (1972).
13. Bowes, G., Ogren, W. L., and Hageman, R. H., Biochem. Biophys. Res. Commun. 45, 716–722 (1971).
14. Bradbeer, J. W., and Anderson, C. M. A., in "The Biochemistry of Chloroplasts" (T. W. Goodwin, ed.), Vol. 2, pp. 175–179. Academic Press, New York, 1967.
15. Bruin, W. J., Nelson, E. B., and Tolbert, N. E., Plant Physiol. 46, 386–391 (1970).

16. Calvin, M., Bassham, J. A., Benson, A. A., Lynch, V. H., Ouellet, C., Schou, L., Stepka, W., and Tolbert, N. E. *Symp. Soc. Exp. Biol.* **5**, 284–305 (1951).
17. Calvin, M., and Massini, P., *Experientia* **8**, 445–457 (1952).
18. Cooper, T. G., Filmer, D., Wishnick, M., and Lane, M. D., *J. Biol. Chem.* **244**, 1081–1083 (1969).
19. Deamer, D. W., and Packer, L., *Biochim. Biophys. Acta* **172**, 539–545 (1969).
20. deDuve, C., *Proc. Roy. Soc., ser. B* **173**, 71–83 (1969).
21. Dilley, R. A., and Vernon, L. P., *Arch. Biochem. Biophys.* **111**, 365–375 (1965).
22. Döhler, G., and Braun, F., *Planta* **98**, 357–361 (1971).
23. Dodd, W. A., and Bidwell, R. G. S., *Plant Physiol.* **47**, 779–783 (1971).
24. Fonseca-Wollheim, F. da, Bock, K. W., and Holzer, H., *Biochem. Biophys. Res. Commun.* **9**, 466–471 (1962).
25. Fortier, N. E., Galland, L., and Lionetti, F. J., *Arch. Biochem. Biophys.* **119**, 69–75 (1967).
26. Gibbs, M., *Ann. N.Y. Acad. Sci.* **168**, 356–368 (1969).
27. Gibbs, M., Ellyard, P. W., and Latzko, E., *in* "Comparative Biochemistry and Biophysics of Photosynthesis," pp. 387–399. Univ. Park Press, State College, Pennsylvania, 1968.
28. Goulding, K. J., Lord, J. M., and Merrett, M. J., *J. Exp. Bot.* **20**, 34–45 (1969).
29. Griffith, T., and Byerrum, R. U., *J. Biol. Chem.* **234**, 762–764 (1959).
30. Hatch, M. D., Osmond, C. B., and Slatyer, R. O., eds., "Photosynthesis and Photorespiration." Wiley (Interscience), New York, 1971.
31. Hess, J. L., and Tolbert, N. E., *J. Biol. Chem.* **241**, 5707–5711 (1966).
32. Hess, J. L., and Tolbert, N. E., *Plant Physiol.* **42**, 371–379 (1967).
33. Hess, J. L., and Tolbert, N. E., *Plant Physiol.* **42**, 1123–1130 (1967).
34. Hess, J. L., Tolbert, N. E., and Pike, L. M., *Planta* **74**, 278–285 (1967).
35. Jackson, W. A., and Volk, R. J., *Annu. Rev. Plant Physiol.* **21**, 385–432 (1970).
36. Jagendorf, A. T., and Hind, G., (1963). *Nat. Acad. Sci.—Nat. Res. Counc., Publ.* **1145**, 599–610.
37. Jimenez, E., Baldwin, R. L., Tolbert, N. E., and Wood, W. A., *Arch. Biochem. Biophys.* **98**, 172–175 (1962).
38. Kearney, P. C., and Tolbert, N. E., *Arch. Biochem. Biophys.* **98**, 164–171 (1962).
39. Kisaki, T., Imai, A., and Tolbert, N. E., *Plant Cell Physiol.* **12**, 267–273 (1971).
40. Kisaki, T., and Tolbert, N. E., *Plant Physiol.* **44**, 242–250 (1969).
41. Kisaki, T., and Tolbert, N. E., *Plant Cell Physiol.* **11**, 247–258 (1970).
42. Kisaki, T., Yano, N., and Hirabayashi, S., *Plant Cell Physiol.* **13**, 581–584 (1972).
43. Lewin, R. A., *Bull. Jap. Soc. Physol.* **5**, 74–75 (1957).
44. Lord, J. M., Codd, G. A., and Merrett, M. J., *Plant Physiol.* **46**, 855–856 (1970).

45. Lorimer, G. H., Ph.D. Thesis, Michigan State University, East Lansing (1972).
46. Lorimer, G. H., Andrews, T. J., and Tolbert, N. E., *Biochemistry* **12**, 18–23 (1973).
47. Mortimer, D. C., *Can. J. Bot.* **39**, 1–5 (1961).
48. Nelson, E. B., and Tolbert, N. E., *Arch. Biochem. Biophys.* **141**, 102–110 (1970).
49. Newmann, J., and Jagendorf, A. T., *Arch. Biochem. Biophys.* **107**, 109–119 (1964).
50. Ogren, W. L., and Bowes, G., *Nature (London) New Biol.*, **230**, 159–160 (1971).
51. Örström, A., *Arch. Biochem. Biophys.* **33**, 484–485 (1951).
52. Orth, G. M., Tolbert, N. E., and Jimenez, E., *Plant Physiol.* **41**, 143–147 (1966).
53. Osmond, C. B., and Harris, B., *Biochim. Biophys. Acta* **234**, 270–282 (1971).
54. Packer, L., Allen, J. M., and Starks, M., *Arch. Biochem. Biophys.* **128**, 142–152 (1968).
55. Pedersen, T. A., Kirk, M., and Bassham, J. A., *Physiol. Plant.* **19**, 219–231 (1966).
56. Peters, J. W., Butz, D. C., and Young, J. W., *J. Dairy Sci.* **54**, 1509–1517 (1971).
57. Plaut, Z., and Gibbs, M., *Plant Physiol.* **45**, 470–474 (1970).
58. Poskuta, J., *Experientia* **24**, 296–297 (1968).
59. Pritchard, G. G., Griffin, W. J., and Whittingham, C. P., *J. Exp. Bot.* **13**, 176–184 (1962).
60. Pritchard, G. G., Whittingham, C. P., and Griffin, W. J., *Nature (London)* **190**, 553–554 (1961).
61. Quayle, J. R., *Biochim. Biophys. Acta* **57**, 398–399 (1962).
62. Rabson, R., Tolbert, N. E., and Kearney, P. C., *Arch. Biochem. Biophys.* **98**, 154–163 (1962).
63. Randall, D. D., and Tolbert, N. E., *J. Biol. Chem.* **246**, 5510–5517 (1971).
64. Randall, D. D., Tolbert, N. E., and Gremel, D., *Plant Physiol.* **48**, 480–487 (1971).
65. Rehfeld, D. W., Randall, D. D., and Tolbert, N. E., *Can. J. Bot.* **48**, 1219–1226 (1970).
66. Rehfeld, D. W., and Tolbert, N. E., *J. Biol. Chem.* **247**, 4803–4811 (1972).
67. Richardson, K. E., and Tolbert, N. E., *J. Biol. Chem.* **236**, 1285–1290 (1961).
68. Romano, M., and Cerra, M., *Biochim. Biophys. Acta* **177**, 421–426 (1969).
69. Rumberg, B., and Siggel, U. *Naturwissenschaften* **56**, 130–132 (1969).
70. Sakami, W., *in* "Amino Acid Metabolism" (W. D. McElroy and B. Glass, eds.), pp. 658–683. Johns Hopkins Press, Baltimore, Maryland, 1955.
71. Sawaki, S., Hattori, N., Morikawa, N., and Yamada, K., *J. Vitaminol.* **13**, 93–99 (1967).
72. Schmid, G. H., *Hoppe-Seyler's Z. Physiol. Chem.* **351**, 575–578 (1970).

73. Schnarrenberger, C., Oeser, A., and Tolbert, N. E., *Plant Physiol.* **48**, 566–574 (1971).
74. Schou, L., Benson, A. A., Bassham, J. A., and Calvin, M., *Physiol. Plant.* **3**, 487–495 (1950).
75. Shain, Y., and Gibbs, M., *Plant Physiol.* **48**, 325–330 (1971).
76. Stafford, H. A., Magaldi, A., and Vennesland. B., *J. Biol. Chem.* **207**, 621–629 (1954).
77. Tanner, H. A., Brown, T. E., Eyster, C., and Treharne, R. W., *Biochem. Biophys. Res. Commun.* **3**, 205–210 (1960).
78. Telfer, A. R., Cammack, R., and Evans, M. C. W., *FEBS Lett.* **10**, 21–24 (1970).
79. Tolbert, N. E., *Nat. Acad. Sci.—Nat. Res. Counc., Publ.* **1145**, 648–662 (1963).
80. Tolbert, N. E., *Annu. Rev. Plant Physiol.* **22**, 45–74 (1971).
81. Tolbert, N. E., *Symp. Soc. Exp. Biol.* **27**, in press (1973).
82. Tolbert, N. E., *in* "Algal Physiology and Biochemistry" (W. D. P. Steward, ed.). Blackwell, Oxford, 1973.
83. Tolbert, N. E., Yamazaki, R. K., and Oeser, A., *J. Biol. Chem.* **245**, 5129–5136 (1970).
84. Tolbert, N. E., and Zill, L. P., *J. Biol. Chem.* **222**, 895–906 (1956).
85. Ullrich, J., *Biochim. Biophys. Acta* **71**, 589–594 (1963).
86. Wang, D., and Burris, R. H., *Plant Physiol.* **38**, 430–439 (1963).
87. Weinhouse, S., *in* "Amino Acid Metabolism" (W. D. McElroy and B. Glass, eds.), pp. 637–657. Johns Hopkins Press, Baltimore, Maryland, 1955.
88. Weinhouse, S., and Friedman, B., *J. Biol. Chem.* **191**, 707–717 (1951).
89. Weissbach, A., and Horecker, B. L., *in* "Amino Acid Metabolism" (W. D. McElroy and B. Glass, eds.), pp. 741–742. Johns Hopkins Press, Baltimore, Maryland, 1955.
90. Whittingham, C. P., Coombs, J., and Marker, A. F. H., *in* "The Biochemistry of Chloroplasts" (T. W. Goodwin, ed.), Vol. 2, pp. 155–173. Academic Press, New York, 1967.
91. Whittingham, C. P., Hiller, R. G., and Bermingham, M., *Nat. Acad. Sci.—Nat. Res. Counc., Publ.* **1145**, 675–683 (1963).
92. Whittingham, C. P., and Pritchard, G. G., *Proc. Roy. Soc., Ser. B* **157**, 366–382 (1963).
93. Wilson, A. T., and Calvin, M., *J. Amer. Chem. Soc.* **77**, 5948–5957 (1955).
94. Wins, P., and Schoffeniels, E., *Biochim. Biophys. Acta* **185**, 287–296 (1969).
95. Wishnick, M., and Lane, M. D., *J. Biol. Chem.* **244**, 55–59 (1969).
96. Wishnick, M., Lane, M. D., Scrutton, M. C., and Mildvan, A. S., *J. Biol. Chem.* **244**, 5761–5763 (1969).
97. Wolfenden, R., *Biochemistry* **9**, 3404–3407 (1970).
98. Young, J. W., Tove, S. B., and Ramsey, H. A., *J. Dairy Sci.* **48**, 1079–1087 (1965).
99. Zelitch, I., *J. Biol. Chem.* **201**, 719–726 (1953).
100. Zelitch, I., *J. Biol. Chem.* **234**, 3077–3081 (1959).

101. Zelitch, I., *Biochem. J.* **77,** 11P (1960).

102. Zelitch, I., *Annu. Rev. Plant Physiol.* **15,** 121–142 (1964).

103. Zelitch, I., *J. Biol. Chem.* **240,** 1869–1876 (1965).

104. Zelitch, I., "Photosynthesis, Photorespiration, and Plant Productivity." Academic Press, New York, 1971.

105. Zelitch, I., and Gotto, A. M., *Biochem. J.* **84,** 541–546 (1962).

Molecular Mechanisms in Blood Coagulation*

Earl W. Davie

*Department of Biochemistry
University of Washington
Seattle, Washington*

Edward P. Kirby

*Department of Biochemistry
Temple University School of Medicine
Philadelphia, Pennsylvania*

I. Introduction

The great clinical relevance of blood coagulation, as well as the drama of seeing a clot appear as if by magic in a tube of blood, has impelled numerous investigators to study the coagulation process. Over the years an abundance of experimental data, some of it useful, has accumulated to help determine the sequences of events

* The unpublished experimental work from this laboratory presented here and the preparation of this article have been supported in part by research grants GM 10793 and HE 11857 from the National Institutes of Health. This review covers material published prior to August, 1972.

and the mechanisms of regulation which are involved in the formation of the fibrin clot. A large number of proteins have been discovered to be necessary for this process, and it is only in recent years that some of these proteins have been isolated and characterized.

The various coagulation factors are listed in Table I. This table includes both the Roman numeral designations for the individual factors and their common names (20). In the past ten years, there has been a general acceptance of the Roman numeral nomenclature for factors V through XIII among investigators working in the field. Fibrinogen, thrombin, prothrombin, calcium ions, and tissue factor, however, are still generally referred to by their common names.

The plasma factors described in Table I are not the only important components affecting hemostasis. *In vivo*, the blood platelets play a primary role in blood coagulation in several ways. Often the first response to damage to a blood vessel is aggregation of platelets at the site of injury. They tend to form a hemostatic plug to inhibit further bleeding. Also, certain platelet components are released during this response of platelets to vessel injury (21). Perhaps most important in coagulation is the release of platelet lipid, termed platelet factor 3, which can function in complex formation with factors IX_a and VIII and also with factors X_a and V. There is also some evidence (22, 23) that platelets may initiate some of the early steps of the intrinsic pathway. Platelets are known to contain factor XIII, and during coagulation platelets also release substances such as serotonin (21, 24) which result in vasoconstriction. The platelets are also important in the retraction of the fibrin clot.

The control and regulation of the blood coagulation mechanism *in vivo* is a complex process involving plasma, platelet, and vascular factors, many of which are very poorly understood. Rather than try to cover all of these, this review will concentrate on the mechanisms of coagulation in the intrinsic and extrinsic systems of mammalian plasma. An attempt will be made to describe on a molecular level the events that occur and how these pathways are regulated. Several good reviews have been written on this topic (25–27). This article will present some of the more recent work, and will try to outline areas that are still only partially understood.

TABLE I

PROPERTIES OF SOME OF THE WELL-DEFINED COAGULATION FACTORS

Coagulation factor		Approximate plasma conc. (mg/1,000 ml)	Coagulation pathway		Molecular weight[b]	Molecular weight references
Roman numeral designation[a]	Common name		Intrinsic system	Extrinsic system		
Factor I	Fibrinogen	2700	+	+	340,000 (human, bovine)	1, 2
Factor II	Prothrombin	140	+	+	68,700 (human); 72,000 (bovine)	3; 4, 5
Factor III	Tissue thromboplastin	—		+	220,000; 330,000 (bovine)	6
Factor IV	Calcium ions	—	+	+	—	
Factor V	Proaccelerin	5–10	+	+	290,000 (bovine)	7
Factor VII	Proconvertin	0.4–0.7		+	63,000 (human)	8
Factor VIII	Antihemophilic factor	15–20	+		1.1 million (human and bovine)	9, 10
Factor IX	Christmas factor	3–5	+		55,400 (bovine)	11
Factor X	Stuart factor	5–10	+	+	55,000 (bovine)	12, 13
Factor XI	Plasma thromboplastin antecedent	—	+		210,000 (human)	14
Factor XII	Hageman factor	—	+		20,000 (human); 82,000 (bovine)	15; 16
Factor XIII	Fibrin stabilizing factor	—	+		300,000 (bovine plasma); 320,000 (human plasma); 146,000 (human platelets)	17; 18, 19; 18, 19

[a] Activated proaccelerin (accelerin) was originally thought to be a new clotting factor and was assigned Roman numeral VI.

[b] Some disagreement exists regarding the exact molecular weight of many of the coagulation factors, as noted in the text. Thus, modifications of the values listed will occur.

II. Molecular Events in Blood Coagulation

Blood coagulation can be divided into two major pathways, referred to as the intrinsic and extrinsic mechanisms (see Table I and Fig. 1). The intrinsic pathway refers to those reactions which lead to fibrin formation and which utilize only factors present in plasma. The extrinsic pathway is composed of plasma factors as well as components present in tissue extracts. Both of these pathways play an important physiological role in mammalian hemostasis. These two mechanisms can be observed separately *in vitro*, but *in vivo* they probably function simultaneously, although to different degrees depending upon conditions. Blood coagulation can also be initiated by other agents, such as Russell's viper venom, but these are not generally of physiological importance.

Some of the coagulation factors participate only in the intrinsic pathway of coagulation, while others participate only in the extrinsic pathway. A number, however, including fibrinogen, factor XIII, prothrombin, calcium ions, factor V, and factor X, participate in both pathways. These two pathways are shown in abbreviated

INTRINSIC PATHWAY

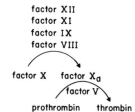

EXTRINSIC PATHWAY

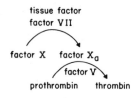

FIG. 1. Abbreviated intrinsic and extrinsic pathways for blood coagulation.

form in Fig. 1. The important point to notice is that factor X is activated independently by the two mechanisms. Thus, it plays a central role in coagulation at a point where the two separate pathways converge.

A. The Intrinsic Activation of Factor X

In 1964, two essentially identical proposals, called the cascade or waterfall mechanisms, were suggested to explain the interaction of the various coagulation factors in the intrinsic pathway of blood coagulation (28, 29). In recent years, these mechanisms have been modified to take into account additional information (Fig. 2). The essential features of the cascade or waterfall concept, however, remain the same. According to the cascade mechanism, various clot-

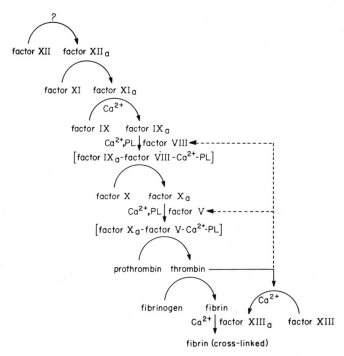

FIG. 2. Tentative mechanism for the initiation of blood clotting in mammalian plasma in the intrinsic system [modified from Davie and Ratnoff (28) and Davie et al. (29)]. PL refers to phospholipid.

ting factors in plasma are present as inactive precursors. When coagulation is initiated, these clotting factors interact with each other in a stepwise manner in which one acts as an enzyme and the other as a substrate. Eventually the interaction of these various factors leads to the formation of thrombin and the fibrin clot. These reactions provide an opportunity for biological amplification in that a few molecules of factor XII_a may activate hundreds of molecules of factor XI within a given time. In turn, the several hundred molecules of factor XI_a may activate thousands of molecules of factor IX. Thus, the level of thrombin which is eventually formed could be many, many times higher than the few molecules of factor XII_a which initiated the coagulation process.

CONTACT ACTIVATION

In the test tube, coagulation is triggered by the activation of factor XII which is converted to factor XII_a by contact with a foreign surface, such as glass (*30–33*). Other surface-active materials, such as kaolin and asbestos, are also effective in the activation of factor XII. The physiological role of factor XII, however, has not been clarified since patients lacking factor XII apparently have no serious coagulation disorders.

Once factor XII_a is formed *in vitro*, it converts factor XI (*34*) to an enzyme (*35–37*). This enzyme, referred to as factor XI_a, appears to be a serine protease since it is sensitive to diisopropylphosphofluoridate (DFP) (*38*). It seems likely that alternate pathways for the activation of factor XI exist, although these reactions have not as yet been defined. The physiological importance of factor XI is well documented since individuals lacking this coagulation factor often have clotting difficulties.

Once factor XI is activated, it in turn activates factor IX (*33, 37, 39–43*). Calcium ions are required for this step. Factor IX is the protein which is inactive in individuals with hemophilia B, a sex-linked coagulation disorder affecting only males. The formation of factor IX_a probably involves proteolysis of the inactive precursor since factors IX and IX_a have different electrophoretic mobilities (*44*) and different elution patterns following gel filtration (*45*). The conversion of factor IX to factor IX_a by proteolysis is consistent with the fact that factor XI_a has esterase activity (*38*). Factor IX_a is also a serine protease (see below) which is inhibited by diisopropyl-phosphofluoridate (*46*). It is not inhibited, however, by soybean trypsin inhibitor (47). Recently, factor IX has been extensively purified (11), and, thus, the exact details of these early steps probably should be clarified soon, since large amounts of pure factor IX free of other clotting factors are now available.

B. Complex Formation between Factor IX$_a$, Factor VIII, Calcium Ions, and Phospholipid

In the next series of reactions, factors IX$_a$ and VIII are involved in the activation of factor X (46, 48–51). Factor VIII is antihemophilic factor, the protein which is inactive in individuals with classic hemophilia. This is another one of the sex-linked clotting disorders affecting only males and is the most common of all the coagulation disorders. The details of these reactions involving factors IX$_a$, VIII, and X are beginning to unfold since highly purified preparations of factors VIII and X have been prepared in recent years.

Factor VIII has been isolated from bovine and human plasma as a large glycoprotein with a molecular weight of approximately 1.2 million (9, 10). It is composed of a number of similar subunits held together by disulfide bonds. The molecular weight of the subunits is approximately 100,000 as determined by sedimentation equilibrium and 240,000 as determined by SDS polyacrylamide gel electrophoresis.

Factor VIII is very sensitive to proteolytic inactivation, and as a result no factor VIII activity is found in serum. It is rather "sticky" protein and tends to bind to surfaces, such as glass, as well as to phospholipid micelles. It does not possess any known esterase or peptidase activity.

The original cascade hypothesis suggested that during coagulation factor VIII was converted to an enzyme, factor VIII$_a$ (28, 29). Factor VIII$_a$ was then thought to convert factor X to factor X$_a$. However, no success has been achieved in various attempts to isolate this postulated factor VIII$_a$ molecule.

More recent evidence suggests that factor VIII acts as a macromolecular cofactor or regulator protein in the conversion of factor X to factor X$_a$ (49, 52). Apparently, a complex is formed between factors IX$_a$ and VIII, phospholipid and calcium ions, and it is this complex which activates factor X (Fig. 3).

Several lines of evidence support the hypothesis that factor VIII acts as a cofactor in a complex between factors VIII, IX$_a$, calcium

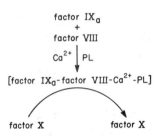

FIG. 3. Complex formation between factors IX_a and VIII and calcium ions and phospholipid (PL) and factor X activation.

ions, and phospholipid. Factor IX_a by itself is active in slowly converting factor X to factor X_a, whereas factor VIII has no known enzymatic activity of its own. Factor VIII, however, dramatically increases the activity of factor IX_a toward its substrate, factor X. The first evidence for the formation of a complex between factors IX_a and VIII was provided by Hougie et al. (49). These investigators showed that the factor X-activating activity of a mixture of factors VIII, IX_a, calcium ions, and phospholipid was eluted at the void volume from a Sephadex G-200 column, suggesting that a complex was formed. Upon dissociating the complex, the activity was lost. The more recent data of Østerud and Rapaport (52) provide additional strong evidence in favor of a complex. In these studies, specific antibodies against factors VIII and IX were employed. When a complex between factors VIII, IX_a, calcium ions, and phospholipid was prepared, it was shown to have the capacity to activate factor X. Addition of antibody against factor IX abolished all capacity to activate factor X, demonstrating that no factor VIII_a-like activity had been formed. Incubation with antibody against factor VIII also abolished activity, showing that factor VIII was necessary for the activation of factor X.

The exact role of factor VIII in this complex is not known, but it might have one or more of the following functions: (a) it may alter the conformation of factor IX_a, increasing the affinity of factor IX_a for factor X, (b) it may alter the conformation of factor X and make the factor X more susceptible to attack by factor IX_a, or (c) it may merely serve to bind factors IX_a and X close together (i.e., concentrating them) and to bring about the correct orientation for factor IX_a to be maximally effective.

A complete understanding of these reactions, however, is complicated by another reaction involving factor VIII. Incubation of this protein with traces of thrombin (0.01–0.1 unit/ml) causes a large increase in the apparent activity of factor VIII as measured in a one-stage clotting assay (53–57). This increase in factor VIII activity is not an artifact due to impurities in the preparations because it is also seen with preparations of highly purified thrombin and factor VIII (10, 58). These preparations do not contain any known contaminating clotting activities. The increase in factor VIII activity is dramatic, approaching nearly 100-fold under the proper conditions. The thrombin-modified factor VIII (factor VIII$_t$) is less stable than the native factor VIII, and its activity decays fairly rapidly.

Several mechanisms have been proposed for this thrombin modification of factor VIII. Biggs et al. (59) originally suggested that factor VIII was associated with fibrinogen and that incubation with low levels of thrombin released it. This now seems unlikely because purified factor VIII which contains no fibrinogen shows a large increase in activity after treatment with thrombin (10). Alternatively, Weiss and Kochwa (60) have proposed that native factor VIII is an oligomer and that incubation with thrombin dissociates it into smaller units which are more active. A third possibility is that thrombin just makes a few breaks in the polypeptide chain(s) of factor VIII to loosen the structure and make it more active without necessarily decreasing the molecular weight.

The precise effect of thrombin modification on the activity of factor VIII is not certain. Østerud et al. (61) have shown that, if a reaction mixture is made between factors IX$_a$, VIII, calcium ions, phospholipid, and factor X, there is a lag in the activation of factor X, and a considerable time is required before significant amounts of factor X$_a$ are generated. However, if factor VIII$_t$ is used instead of factor VIII, factor X activation starts immediately. They interpreted the lag as being the time required for small amounts of thrombin to be generated from traces of prothrombin in their reagents and for this thrombin to modify the factor VIII. In agreement with Rapaport et al. (55), they concluded that native factor VIII was essentially inactive and thrombin modification of the factor VIII was necessary for the formation of an active complex. This hypothesis was supported by the finding that hirudin, an inhibitor of thrombin, could significantly prolong the lag phase.

These findings could explain the apparent increase in factor VIII activity upon thrombin modification as seen in a one-stage assay. When factor VIII_t is added to an assay plasma in the presence of calcium ions and a contact activator, such as koalin, factor VIII_t will rapidly form an active complex with factor IX_a. This leads to an immediate factor X activation. However, when factor VIII has not been thrombin-modified, the formation of a highly active factor $VIII–IX_a$ complex is delayed. Whether factor VIII which has not been modified by thrombin has biological activity remains to be clarified.

The requirements for thrombin modification of factor VIII for it to have maximal activity raises the question of where the necessary thrombin comes from. There are at least four possible explanations: (a) A small amount of tissue factor is released from cells upon injury, and this is able to generate the first traces of thrombin via extrinsic activation of factor X. (b) Factor IX_a by itself or the complex formed between native factor VIII and factor IX_a has partial activity against factor X, resulting in the formation of some factor X_a. This, in turn, would generate the first traces of thrombin necessary to modify factor VIII. (c) Plasma may contain other enzymes, in addition to thrombin, which are capable of modifying factor VIII. (d) There is a constant generation of low levels of thrombin *in vivo*, so that at any one time a certain small percentage of factor VIII is present in the form of factor VIII_t. Thus, a small amount of the active complex could form immediately after the formation of factor IX_a. The existence of a certain fraction of factor VIII in plasma as factor VIII_t, with its resulting higher rate of decay, might explain the high rate of factor VIII turnover *in vivo*. Weiss and Kochwa (60) have indicated that factor VIII_t may have a lower molecular weight than factor VIII, and they observed the presence of some of the lower molecular weight species when plasma was centrifuged, suggesting the presence of factor VIII_t *in vivo*. Since factor VIII_t may have different solubility and absorption properties than native factor VIII (60), it may be selected against by most purification schemes. Niemetz and Nossel (62) have observed an increased factor VIII activity during certain hypercoagulable states and have attributed this to increased levels of factor VIII$_t$ in these patients.

C. The Extrinsic Activation of Factor X

Exposure of blood to extracts of various tissues results in the rapid formation of clotting activity. The material present in these extracts that is responsible for this activity has been termed tissue thromboplastin. Trauma to a tissue may result in the release of this material from the cells and eventually it may come in contact with blood which also has been released at the site of injury. This interaction will lead to the formation of a blood clot.

Tissue thromboplastin has been shown by Nemerson and Spaet (63) to consist of two components, phospholipid and protein. The phospholipid fraction can be replaced by mixtures of synthetic phospholipids (64, 65). The protein fraction, which has been termed tissue factor, contains two components which are very similar but differ in their molecular weights (220,000 and 330,000). Nemerson and Pitlick (6) feel that these are the same protein which differ only in their content of residual-bound phospholipid. Tissue thromboplastin has a membranous appearance in the electron microscope (66, 67), presumably due to its high phospholipid content. Recently, Zeldis et al. (68) have reported that it is present in high concentration in the plasma membrane of endothelial cells. Tissue factor also contains significant peptidase activity (69), but tissue thromboplastin cannot by itself activate factor X. A plasma protein, factor VII, is also required for the activation of factor X by tissue thromboplastin (70).

Factor VII is a plasma protein which is in many ways similar to prothrombin and factors IX and X. It is readily adsorbed by BaSO$_4$ and eluted from DEAE-cellulose at a position similar to prothrombin and factors IX and X. As with factors IX, X, and prothrombin, vitamin K is required for its synthesis. Factor VII has been purified approximately 10,000-fold by Gladhaug and Prydz (71). When isolated from plasma, it has a molecular weight of approximately 59,000, whereas factor VII isolated from serum has a molecular weight of only 49,000. Johnston and Hjort (72) have shown that during the intrinsic coagulation of blood there is an apparent 2- to 3-fold increase in factor VII activity. Prydz (73) observed a 7.5-fold increase. Johnston and Hjort (*72*) observed that the factors involved in the generation of intrinsic factor X activator (factors XII, XI, IX, VIII, and calcium ions) were required for this increase in factor VII activity, but that factors X and V, thrombin, and platelets were not necessary. Altman and Hemker (*74*) have suggested that factor XII_a is responsible for this increase in activity.

Both factor VII and tissue thromboplastin are required for generation of the extrinsic factor X activator, but the exact nature of their interaction is unclear. Two mechanisms have been proposed.

Mechanism a. Tissue thromboplastin activates factor VII to form an enzyme which in turn activates factor X (Fig. 4A). This mechanism postulates the formation of a factor VII_a molecule which can activate factor X by itself in the absence of tissue thromboplastin. Numerous investigators (*12, 70, 75*) have attempted to generate factor VII_a from factor VII and tissue thromboplastin and then purify the factor VII_a, but so far none have been successful. Williams and Norris (*70*) did, however, obtain one preparation of factor VII which spontaneously activated and was able to convert factor X to factor X_a in the absence of added tissue thromboplastin. Addition of lung microsomes (a crude tissue thromboplastin) did not increase this activity. These data might be rationalized on the basis that their preparations of factor VII were almost certainly contaminated with factor IX. Spontaneous activation of the factor IX would result in a factor VII_a-like activity, although the lack of effect of the lung

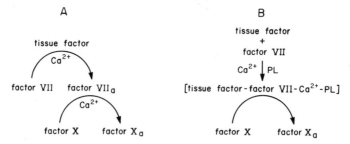

Fɪɢ. 4. Possible mechanisms for the activation of factor X in the extrinsic pathway. (A) Stepwise activation of factor VII and factor X. (B) Complex formation between tissue factor, factor VII, calcium ions, and phospholipid (PL).

microsomes would require a simultaneous inactivation of all the true factor VII activity.

Recently, Østerud et al. (12) have reported obtaining factor VII$_a$ activity by activating factor VII with tissue thromboplastin and then destroying the tissue thromboplastin activity with phospholipase C. The resulting factor VII$_a$ preparation still contained phospholipid (further incubation with phospholipase C destroyed the factor VII$_a$ activity). Unfortunately, they could not assay for tissue thromboplastin activity in the presence of factor VII$_a$ so the demonstration that the factor VII$_a$ preparation contained no tissue thromboplastin had to be based on the inactivation of tissue thromboplastin in parallel incubations of tissue thromboplastin with phospholipase C in the absence of factor VII.

Mechanism b. A second possibility, suggested by Nemerson (75), is that factor VII and tissue factor form a complex in the presence of calcium ions and phospholipid, and it is this complex which is the extrinsic factor X activator (Fig. 4B). Nemerson (75) has shown that neither factor VII nor tissue factor is sensitive to DFP or to soybean trypsin inhibitor, but upon formation of the complex the resulting factor X activator activity is sensitive to both DFP and soybean trypsin inhibitor. Pitlick et al. (69) have shown that tissue factor by itself has peptidase activity, but it is not known whether this activity is directed against factor VII or X. Presumably, tissue factor could activate factor VII to factor VII$_a$ which is only active while bound to the complex. Alternatively, tissue factor might act on factor X in a manner analogous to factor IX$_a$ or Russell's viper venom, whereas factor VII acts only like a cofactor, analogous to factor VIII or V.

D. The Mechanism of Factor X Activation

Factor X is a plasma glycoprotein containing approximately 10% carbohydrate. It can readily be purified free of prothrombin, factor IX, and factor VII by chromatography on DEAE-cellulose (13, 76). There is disagreement in the literature about the molecular weight of plasma factor X. Esnouf and Williams (77) and Seegers et al. (78) obtained a molecular weight of approximately 86,000, while Jackson and Hanahan (13) and Fujikawa et al. (76) obtained a value of 55,000. It appears probable that the molecular weight estimate of 86,000 is due to minor aggregation of the protein during the molecular weight determination. According to Esnouf and Williams (77), factor X obtained from serum has a molecular weight of 36,000, presumably due to limited proteolysis by some of the enzymes generated during clotting.

Bovine factor X can be separated into two biologically active fractions, called factors X_1 and X_2, by column chromatography on DEAE-Sephadex (13, 76). Each has a molecular weight of approximately 55,000, and their amino acid and carbohydrate compositions appear to be essentially the same (76). Recently, however, Jackson (79) has noted slight differences in the carbohydrate compositions of the two proteins.

After reduction with 2-mercaptoethanol, bovine factor X can be separated into a heavy and a light chain with molecular weights of 38,000 and 17,000, respectively (76). The heavy chains of both factors X_1 and X_2 have an amino terminal sequence of Trp-Ala-Ile-His-, and the light chains have an amino terminal sequence of Ala-Asn-Ser-Phe-. Essentially all the carbohydrate found in factor X is located in the heavy chain of the molecule.

As previously mentioned, factor X can be converted to factor X_a in a variety of ways: (a) by incubation with the intrinsic complex made up of factors IX_a and VIII, calcium ions and phospholipid, (b) by incubation with the extrinsic activator composed of tissue factor, factor VII, calcium ions and phospholipid, or (c) by incubation with certain proteolytic enzymes, such as trypsin or a protease isolated from Russell's viper venom. Some of the details of these reactions have recently been clarified by Fujikawa et al. (80), who studied the activation of bovine factor X by trypsin and the protease from Russell's viper venom.

It was found that during the activation reaction, Russell's viper venom splits a glycopeptide with a molecular weight of approximately 11,000 from the amino terminal end of the heavy chain. This decreases the molecular weight of the precursor molecule from 55,000 to 44,000 and results in the formation of a new amino terminal sequence of Ile-Val-Gly-Gly- in the heavy chain of factor X_{1a}. No change in the amino terminal end of the light chain in factor X_1 occurs during the activation reaction. These reactions are summarized in Fig. 5.

A similar situation exists in the activation of factor X by trypsin (81). However, in this case trypsin also hydrolyzes a polypeptide

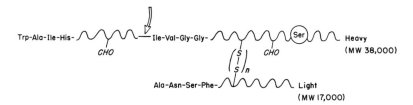

Fig. 5. Structure of factor X_1 showing the peptide bond split during activation by a protein from Russell's viper venom. From Fujikawa *et al.* (*80*).

of approximately 3000 molecular weight from the carboxyl terminal end of the heavy chain. This yeilds a factor X_a with a molecular weight of approximately 41,000. This is slightly smaller than that formed by the protease from Russell's viper venom. Again, apparently no change occurs in the light chain during activation by trypsin.

The active site serine is located in the heavy chain of factor X_a, and recent evidence (*82*) indicates that there is considerable homology in the amino acid sequence of the active site region of factor X_a with other serine proteases (Table II). A similar situation exists for the amino terminal portion of factor X_{1a}, as shown in Table III. The amino acids which are underlined in Tables II and III are identical to those found in the amino terminal region and active site region of bovine trypsin, indicating considerable homology between the two proteins. The amino acid sequence in these two regions is also very similar to other proteases, such as bovine thrombin, bovine chymotrypsin, human plasmin, and porcine elastase. It is clear from this evidence that factor X_a is a typical serine protease with numerous regions of homology common to other pancreatic and proteolytic enzymes. Although the details for the activation of factor X by the intrinsic and extrinsic systems have not been established, it appears extremely likely that the activation reaction by these mechanisms will involve the split of the same amino terminal bond of the heavy chain of factor X, giving rise to the Ile-Val-Gly-Gly- sequence. Thus, these data strongly suggest that the complex of factors IX_a and VIII, as well as the tissue factor–factor VII complex, possess protease activity with a specificity like trypsin for basic amino acids, such as arginine.

TABLE II

AMINO TERMINAL SEQUENCES OF FRAGMENT C-2 OF THE HEAVY CHAIN OF FACTOR X_{1_a}
AND OF SEVERAL OTHER PROTEOLYTIC ENZYMES

Enzymes	Amino terminal sequence[a]
	5 10
Bovine factor X_{1_a} (fragment C-2)	Ile-Val-Gly-Gly-Arg-Asp-Cys-Ala-Glu-Gly-Glu-Glu-Cys-
Bovine trypsin[b]	Ile-Val-Gly-Gly-Tyr-Thr-Cys-Gly-Ala-Asn-Thr-Val-
Bovine thrombin[c]	Ile-Val-Glu-Gly-Gln-Asp-Ala-Glu-Val-Gly-Leu-Ser-
Bovine chymotrypsin A[d]	Ile-Val-Asn-Gly-Glu-Glu-Ala-Val-Pro-Gly-Ser-Trp-
Human plasmin[e]	Val-Val-Gly-Gly-Gln-Val-Ala-His-Pro-His-Ser-Trp-
	15 20 25
Bovine factor X_{1_a}	-Pro-Trp-Gln-Ala-Leu-Leu-Val-Asn-Glu-Glu-Asn-Glu-Gly-Gly-
Bovine trypsin	-Pro-Tyr-Gln-Val-Ser-Leu-Asn-Ser-Gly-Tyr-His-*/*[f]
Bovine thrombin	-Pro-Trp-Gln-Val-Met-Leu-Phe-Arg-Lys-Ser-Pro-Gln-Glu-Leu-[g]
Bovine chymotrypsin A	-Pro-Trp-Gln-Val-Ser-Leu-Gln-Asp-Lys-Thr-Gly-Phe-His-
Human plasmin	-Pro-Trp-Gln-Val-Val-Leu-Arg-
	30 35
Bovine factor X_{1_a}	-Phe-Cys-Gly-Gly-Thr-Ile-Leu-Asn-Glu-Phe-Tyr-Val-
Bovine trypsin	-Phe-Cys-Gly-Gly-Ser-Leu-Ile-Asn-Ser-Gln-Trp-Val-
Bovine thrombin	-Leu-Cys-Gly-Ala-Ser-Leu-Ile-Ser-Asp-Arg-Trp-Val-
Bovine chymotrypsin A	-Phe-Cys-Gly-Gly-Ser-Leu-Ile-Asn-Glu-Asn-Trp-Val-

[a] Taken from Titani et al. (82). The underscored amino acids are homologous with the heavy chain of bovine factor X_{1_a}.
[b] Walsh and Neurath (83).
[c] B chain of thrombin (84).
[d] B chain of α-chymotrypsin (85).
[e] B chain of plasmin (86).
[f] Deletions in bovine trypsin correspond to residues 24 and 25 of factor X_{1_a}.
[g] Leucine represents an apparent insertion in bovine thrombin between residues 25 and 26 of factor X_{1_a}.

TABLE III

Amino Acid Sequences of the Sites that React with Diisopropyl-phosphofluoridate in the Heavy Chain of Bovine Factor X_{1a} and in Other Serine Proteases

Enzyme	Amino acid sequence[a]
	170 175
Bovine factor X_1[c]	Phe-Cys-Ala-Gly-Tyr-Asp-Thr-Gln-Pro-Glu-
Bovine trypsin[e]	Phe-Cys-Ala-Gly-Tyr-Leu-Glu-Gly-Gly-Lys-
Bovine thrombin[i]	Phe-Cys-Ala-Gly-Tyr-Lys-Pro-Gly-Glu-Gly-Lys-[d]Arg-[d]Gly-[d]
Bovine chymotrypsin A[e]	Ile-Cys-Ala-Gly-*/- Ala-Ser-Gly-Val-*[g]
Porcine elastase[h]	Val-Cys-Ala-Gly-*/- Gly-Asn-Gly-Val-Arg-
	180 185 190
Bovine factor X_1[c]	-Asp-Ala-Cys-Gln-Gly-Asp-SER-Gly-Gly-Pro-His-Val-Thr-Arg-
Bovine trypsin	-Asp-Ser-Cys-Gln-Gly-Asp-SER-Gly-Gly-Pro-Val-Val-Cys-Ser-
Bovine thrombin	-Asp-Ala-Cys-Glu-Gly-Asp-SER-Gly-Gly-Pro-Phe-Val-Met-Lys-
Bovine chymotrypsin A	-Ser-Ser-Cys-Met-Gly-Asp-SER-Gly-Gly-Pro-Leu-Val-Cys-Lys-
Porcine elastase	-Ser-Gly-Cys-Gln-Gly-Asp-SER-Gly-Gly-Pro-Leu-His-Cys-Leu-

[a] Taken from Titani et al. (82). The underscored amino acids are homologous with the heavy chain of bovine factor X_{1a}. The numbers refer to the amino acid sequence of bovine trypsin (83).

[b] Walsh and Neurath (83).

[c] B chain of thrombin (84).

[d] Lysine, arginine, and glycine are apparent insertions in bovine thrombin between residues 176 and 177 of bovine trypsin.

[e] B chain of α-chymotrypsin (85).

[f] Deletions in bovine chymotrypsin and porcine elastase correspond to residue 171 of bovine trypsin.

[g] Deletion in bovine chymotrypsin corresponds to residue 176 of bovine trypsin.

[h] Hartley and Shotton (87).

E. Complex Formation between Factor X_a, Factor V, Calcium Ions, and Phospholipid

Factor V has been isolated and characterized by several groups (*7, 88, 89*). It is a large glycoprotein, with a molecular weight of about 290,000 (*7*). Factor V acts as a cofactor for factor X_a in the formation of the "prothrombinase" complex (Fig. 6). Philip *et al.* (*90*) have indicated that factor V may exist in plasma in more than one form, due to dissociation or complexing with lipid.

Incubation of factor V with low concentrations of thrombin leads to an increase in the specific activity of factor V. The extent of this increase varies with the factor V preparation. Papahadjopoulos *et al.* (*88*) and Newcomb and Hoshida (*91*) have reported a 10-fold increase, Barton and Hanahan (*92*) and Colman (*89*) a 2- to 3-fold increase, whereas Jobin and Esnouf (*93*) stated that no increase in activity occurs. At least part of this discrepancy may be due to degradation of factor V which occurs when it is purified on cellulose phosphate, as in the procedure for Esnouf and Jobin (*7*). Newcomb and Hoshida (*91*) indicated that the extent of increase seems to be inversely related to the original specific activity of the factor V.

Thrombin modification of factor V causes it to be more unstable. This instability is apparently not due to continued degradation by thrombin, because the rate of decay of the activity is not dependent on the thrombin concentration. Addition of hirudin, which blocks the activity of thrombin, does not alter the rate of decay seen after thrombin modification (*89*).

The mechanism of thrombin modification of factor V is uncertain.

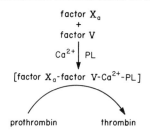

Fig. 6. Complex formation between factor X_a, factor V, calcium ions, and phospholipid (PL), and the activation of prothrombin.

It presumably involves a limited proteolysis of the protein, because a similar increase in activity can be obtained with an esterase isolated from Russell's viper venom (94, 95) or with papain (96). Colman et al. (97) reported that, when factor V is thrombin-modified, no change in molecular weight (as measured by gel filtration) is observed. Papahadjopoulos et al. (88), however, reported a significant decrease in the molecular weight of factor V upon incubation with thrombin.

The kinetics of thrombin modification are somewhat complicated, and the amount of activation is dependent on the concentrations of both factor V and thrombin. Both product and substrate inhibition seem to be involved (89). Addition of factor V_t to a mixture of factor V and thrombin can block further activation. The factor V_t apparently inhibits by binding to unmodified factor V because the esterase and fibrinogen-clotting activities of the thrombin are not inhibited. Thrombin modification of factor V seems to require a higher concentration of thrombin (1 unit/ml) than is required for modification of factor VIII (0.01–0.1 unit/ml).

It was originally proposed (98) that activated factor X in turn activated factor V to form a new enzyme, factor V_a, which was responsible for the conversion of prothrombin to thrombin. An alternative hypothesis is that factor X_a has the enzymatic site which converts prothrombin to thrombin, and that factor V acts in some way as a high molecular weight cofactor or regulator protein, analogous to factor VIII and its interaction with factor IX_a.

Most evidence seems to suggest that it is the second hypothesis which is correct. Factor X_a can convert prothrombin to thrombin in the absence of factor V (93), suggesting that it is factor X_a which has the active site. The presence of factor V increases the rate of prothrombin conversion severalfold (93, 99), but no evidence has been found for the existence of factor V_a (93). If an active "prothrombinase" complex is made between factors X_a, V, calcium ions, and phospholipid, and then it is dissociated with ethylenediaminetetraacetate (EDTA), factor V can be recovered from this mixture unchanged (100), suggesting that it is not converted to factor V_a. The kinetic data of Hemker (101) also appear to be inconsistent with a mechanism involving production of factor V_a, which is active by itself on prothrombin.

Phospholipid has been shown to be necessary for full activity of the factor X_a–factor V complex (93). Both factors X_a and V bind to phospholipid micelles at physiological calcium ions concentrations (102). Factor X_a and factor V must be bound to the same phospholipid micelle for optimal activity (103). The substrate of the reaction, prothrombin, also binds to phospholipid, but the product of the reaction, thrombin, does not (104). Desorption of the thrombin from the phospholipid micelle would thus tend to pull the reaction in the forward direction.

The "prothrombinase" complex involves stoichiometric amounts of factors X_a and V (104), but the exact composition and structure of this complex has not been determined. The data of Papahadjopoulos et al. (88) suggest that factor X_a has binding sites for both factor V and phospholipid, and that calcium ions are required for this binding.

The exact role of factor V in the complex is not known. It may just act by binding factor X_a and prothrombin in proximity to one another, resulting in an effective concentration of the reactants with a resulting increase in the rate of prothrombin conversion. Alternatively, it might alter the conformation of prothrombin in some way to increase its susceptibility to factor X_a. A third possibility is that binding of factor X_a to factor V in some way increases the factor X_a activity. Jobin and Esnouf (93) did not observe a significant increase in the esterase activity of factor X_a when added to factor V, calcium ions, and phospholipid. In contrast, Colman (99) reported a 3-fold increase of the esterase activity of factor X_a against tosylarginine methyl ester (TAME) when factor V, phospholipid, and calcium ions were added. Moreover, Colman (99) observed a change in the relative activity of factor X_a on a series of different synthetic esters when factor V was added.

Many of the earlier studies of the interaction of factors V and X_a were complicated by the fact that crude Russell's viper venom was used to activate the factor X, and this venom also contains a protein capable of increasing the activity of factor V (94, 105) in a manner similar to thrombin modification.

Thrombin modification of factor V is apparently not essential for its activity, although Prentice (106) reported that, when the complex is formed with factor V which has not been thrombin modified, there is a lag in the generation of thrombin from prothrombin. Incubation in the presence of preformed thrombin abolished this lag. This is very similar to the effect of thrombin on factor VIII in the formation of the factor IX_a–VIII complex.

F. Prothrombin Activation

Conversion of prothrombin to thrombin is accompanied by a decrease in molecular weight from 72,000 to 39,000 or less, and the appearance of enzymatic activity (4).

Prothrombin activation is not a simple one-step process, but it involves a series of changes in the prothrombin molecule, apparently due to the protease activity of the factor X_a–factor V complex. Intermediates exist which have a molecular weight different from that of prothrombin, suggesting that they are formed by limited proteolysis. These intermediates, however, do not have esterase or fibrinogen-clotting activity (4). Seegers et al. (107) state that 2 moles of thrombin are generated from each mole of prothrombin, but Magnusson (108) claims that only 1 mole of thrombin is formed, with the rest of the prothrombin molecule yielding a second esterase which is different from thrombin. The data of Mann et al. (4) also indicate that only 1 mole of thrombin is formed, but these investigators stated that the remaining peptide which is cleaved from the prothrombin is inactive. They also suggest that the first step in prothrombin activation is the hydrolysis of a small fragment to yield an intermediate with a molecular weight of approximately 65,000 followed by the splitting of this molecule into two pieces with molecular weights of 39,000 and 24,000. Neither of these fragments has esterase activity. The 24,000 molecular weight fragment does not yield any activity upon further digestion, but subsequent proteolysis of the 39,000 molecular weight fragment generates a series of thrombin molecules which differ in their molecular weights, but all have esterase and fibrinogen-clotting activity.

Magnusson (84, 109) has determined the amino acid sequences of the A and B chains of the 34,000 molecular weight bovine thrombin. The A chain contains 49 residues with an N-terminal threonine and a C-terminal arginine. The B chain contains 265 residues with an N-terminal isoleucine and a C-terminal arginine. This chain also contains the active site serine and shows considerable homology to factor X_{1a} (see Tables II and III). These results are consistent with the concept that many of the coagulation factors which are synthesized in the liver have evolved from a common ancestor in a manner similar to the evolution of the pancreatic proteases (87).

G. Fibrin Formation and the Cross-Linking Reaction

Fibrinogen is readily attacked by thrombin, resulting in the liberation of fibrinopeptides A and B and the formation of the fibrin monomer (110, 111). A great deal of information has been obtained in recent years regarding the structure of fibrinopeptides from various mammalian and avian species (112–117). The fibrinopeptides are released from the amino terminal end of fibrinogen by hydrolysis of specific arginyl-glycyl bonds by thrombin. After release of the fibrinopeptides, the resulting fibrin monomers undergo polymerization to form a fibrin clot. Herzig et al. (118) have shown that it is the release of a critical amount of fibrinopeptide A which is the crucial step in the initiation of polymerization. The resulting fibrin clot is readily soluble in denaturing agents and is very susceptible to attack by degradative enzymes, such as plasmin. In the presence of factor XIII_a and calcium ions, the soluble fibrin clot is then cross-linked to form a tough, insoluble fibrin.

Factor XIII is activated by thrombin in the presence of calcium ions (119–121). This reaction has been studied independently in a number of laboratories (122–128). Factor XIII isolated from human plasma has a molecular weight of 320,000. It is composed of two a subunits, each with a molecular weight of approximately 75,000, and two b subunits, each with a molecular weight of approximately 88,000 (18, 19).

The a and b subunits are held together by noncovalent bonds since they are readily separated in the presence of sodium dodecyl sulfate and in the absence of a reducing agent, such as 2-mercaptoethanol. When factor XIII is isolated from human platelets, however, the protein has a molecular weight of 146,000 and is composed of two a chains of 75,000 molecular weight. This molecule has no b subunits. The a subunits of plasma factor XIII appear to be identical with the a subunits of platelet factor XIII. The a subunits from plasma or platelet factor XIII do not contain carbohydrate and both are rapidly modified by thrombin during the activation reaction. In this reaction, the a subunits of factor XIII are reduced in molecular weight by approximately 4000. Apparently, there is no change in the b subunits of plasma factor XIII during the activation reaction.

After activation, factor XIII_a rapidly forms cross-linkages between the γ chains of two different fibrin monomers. In this reaction the ϵ-amino group from lysine from one fibrin molecule forms a peptide linkage with the glutaminyl residue of an adjacent fibrin monomer. On longer incubation, cross-linkage also occurs between the α chains of different fibrin monomers, leading to high molecular weight polymers of the α chains. Evidence that the γ-γ linkages are intermolecular as opposed to intramolecular was obtained by isolating the donor acceptor pairs from mixed human and bovine fibrin preparations (129). The total number of cross-linkages appears to be about six per mole of fibrin (130), four of which involve cross-linkages with the α chains and two with the γ chains (131). After the cross-linking reactions, the fibrin clot is much more resistant to lysis (132).

III. Regulatory Mechanisms in Blood Coagulation

Perhaps the most striking feature of blood coagulation is the almost explosive generation of thrombin after a brief lag period in which little or no thrombin is produced. Although large amounts of thrombin are produced from an apparently rather small stimulus, under normal conditions *in vivo* there is little or no tendency to form clots spontaneously.

Very sensitive regulation can occur in the clotting system because: (a) the enzymes involved are linked together in sequence, (b) most of the reactions occur on surfaces rather than free in solution, and (c) there is feedback regulation at several stages in the coagulation mechanism.

1. LINKED ENZYMES

The cascade mechanism can provide an explanation for the rapid generation of thrombin from a very small stimulus because it suggests that the system can function as a multistage amplifier. For each molecule of factor XII which is activated by surface contact, several molecules of factor XI will be activated. Each of these molecules of factor XI_a will in turn generate many molecules of factor IX_a, and so on down the scheme.

Several requirements must be met to produce an efficient amplification. Increasing amounts of clotting factors must be available as one proceeds from factor XII down to prothrombin. In general, this is true (see Table I). The specific activities of those clotting factors which participate as enzymes must be fairly comparable. Also, there must be no "bottlenecks" or rate-limiting reactions at later stages.

Amplifications may be occurring at several stages of coagulation. For instance, during clotting *in vitro* in the intrinsic pathway, essentially all of the prothrombin (which is approximately 2 μM in the plasma, see Table I) is converted to thrombin. Of the factor X (approximately 0.2 μM in the plasma) only 10–20% is activated during clotting. Thus, each molecule of factor X_a which is generated (i.e., approximately 0.04 μM) activates on the average approximately 50 molecules of prothrombin. The 2 μM thrombin which results causes the conversion of essentially all the fibrinogen (approximately 10 μM) to fibrin. Actually, this is probably a great underestimate of the amplification involved, because it has been observed (80) that as little as 1 to 2 $\times$ 10^{-9} g of purified factor X_a can clot a standard assay mixture of factor X-deficient plasma (i.e., containing approximately 4 $\times$ 10^{-4} g of fibrinogen) in about 30 seconds. This amount of factor X_a is about 0.1% of the factor X found in 0.1 ml of normal plasma. When higher levels of factor X_a are used, comparable to levels that would be generated in plasma during extrinsic coagulation, clotting occurs in less than 10 seconds. This suggests that it is probably the early stages of the coagulation mechanism which are rate limiting.

The analogy between the intrinsic pathway and a simple amplifier, however, is not truly justified because at several stages macromolecular cofactors (such as factors V and VIII) are required. These factors tend to regulate the activities of factor X_a and IX_a, respectively, and probably lead to "bottlenecks" in the coagulation mechanism.

2. FEEDBACK MECHANISMS

Positive feedback in the amplifier seems to be important in the blood coagulation scheme. Factor VIII in its native state appears to form an active factor IX_a–VIII complex rather slowly (*55*). Traces of thrombin, however, cause a large increase in factor VIII activity by modifying factor VIII in such a way that it can form an active complex with factor IX_a.

The low activity of native factor VIII presumably "cuts off" the cascade at the level of the factor IX_a–VIII complex, causing a pile-up of factor IX_a. However, modification of factor VIII by the first traces of thrombin causes a rapid formation of an active IX_a–VIII_t complex, with the resulting rapid generation of factor X_a.

A similar situation occurs with the factor X_a–V complex, in that low concentrations of thrombin greatly increase the activity of factor V. Thus, the generation of thrombin by the intrinsic pathway is quite slow and inefficient until the first traces of thrombin are formed. This small amount of thrombin then modifies factors VIII and V, greatly increasing the efficiency of the intrinsic pathway and causing rapid generation of large amounts of thrombin. These effects are illustrated in Fig. 2.

When factor VIII-deficient plasma is recalcified, it is frequently observed that a few strands of fibrin are formed within the normal time, but the subsequent fibrin formation is very slow, rather than the rapid generation seen in normal plasma. This suggests that the hemophilic plasma has the capacity to generate the first traces of thrombin. Since factor VIII is not present, however, an active factor IX_a–VIII_t complex cannot be formed and the "bottleneck" remains. Thus, coagulation proceeds at a slow pace.

3. REACTIONS AT SURFACES AND THE ROLE OF PHOSPHOLIPID

It is striking that several of the major reactions in the pathway of blood coagulation seem to occur when the proteins involved are attached to surfaces. Factor XII is activated by contact with a surface and probably activates factor XI while still bound to the surface. Phospholipid micelles are required for both the extrinsic and intrinsic factor X-activating complexes as well as for maximal activity of the factor X_a–factor V complex.

The simplest explanation of the role of the phospholipid is that it acts to bind the various clotting factors at its surface and in this way increases the effective concentration of these factors. It can be seen from Table I that the concentration of most of the coagulation factors in the plasma is quite low. If these were uniformly distributed throughout the plasma and factor number 1 had to diffuse into the vicinity of factor number 2 before the two could interact, the rate of the reaction would be very slow, even if the two proteins reacted instantaneously upon contact. The enzymes are not held rigidly on the phospholipid but probably "float" on the surface and can diffuse toward one another to interact.

In connection with this, the studies of Mattiasson and Mosbach (*133*) on linked enzyme systems bound to an insoluble matrix are of interest. They studied the following series of reactions:

$$
\begin{array}{ccccc}
 & \beta\text{-galac-} & & & \text{Glu-6-P} \\
 & \text{tosidase} & \text{hexokinase} & & \text{dehydrogenase} \\
\text{Lactose} & \longrightarrow & \text{Glucose} \longrightarrow & \text{Glu-6-P} \longrightarrow & \text{6-P-Gluconate} \\
 & & \text{ATP} \quad \text{ADP} & & \text{NAD} \quad \text{NADH}
\end{array}
$$

and observed that when the enzymes were free in solution there was an appreciable lag before significant amounts of NADH would appear, reflecting the time required for the necessary intermediates at the earlier stages to reach significant concentrations. When the three enzymes were bound to a matrix, however, NADH appeared more rapidly indicating that the intermediates were efficiently transferred from one enzyme to another, rather than diffusing out into the medium. One would expect this effect to be even more pronounced where the "substrates" are large, slowly diffusing macromolecules.

At the concentration of phospholipid used for most coagulation assays (i.e., 1 mM) the phospholipid can be expected to be entirely in the form of micelles (the critical concentration for micelle formation by these lipids is on the order of 10^{-9} M or less) (*134*). These micelles are presumably bimolecular leaflets (*135*) whose size is dependent upon the amount of phospholipid present and the previous treatment of the solution, in terms of the amount of shear, freezing, etc., that it has been subjected to. The experiments of several workers (*136–138*) have suggested that it is the charge on the phospholipid micelle which is the important factor in determining its efficiency in blood coagulation, although the size and structure of the phospholipid molecule may play a role (*135*). Davie et al. (*26*) reported that mixtures of phosphatidylserine and phosphatidylcholine or phosphatidylethanolamine had high activity in the formation of the factor IX–VIII complex, although singly these phospholipids had little or no activity. Mixtures of other phospholipids also had considerable activity. Similar phospholipid requirements had earlier been demonstrated for the factor X_a–factor V complex (137).

Adsorption of the relevant coagulation factors to the phospholipid micelles in effect serves to concentrate them from the plasma and so promote their interaction. Factors II, V, VIII, IX, and X have all been shown to bind to phospholipid at physiological concentrations of calcium ions (102, 139). The coagulation factors are not bound rigidly to the micelle surface, but can exchange bonds with various phospholipid molecules and so migrate across the surface. The phospholipid molecules themselves are also not held in position but are free to diffuse into optimal contact with one another.

Whether the phospholipid micelle serves as more than just a surface for congregation of the coagulation factors is not certain. Adsorption of a protein to a surface may well produce a profound change in the conformation of the protein, altering either its specificity, or, in the case of a zymogen, its susceptibility to proteolytic attack.

Schmer et al. (9) made the interesting observation that factor VIII would readily bind to glass, and still maintain its activity while bound to the glass surface. Moreover, when bound to glass, the requirement for phospholipid was relieved. In other words, the glass surface had the same effect on the factor VIII as binding to a phospholipid micelle.

Prothrombin has been shown to bind very well to phospholipid, but thrombin does not bind. After conversion of prothrombin to thrombin, the thrombin could be expected to diffuse away from the surface of the micelle, leaving the region adjacent to the factor X_a–factor V site free for more prothrombin.

Although phospholipid plays an important role at several stages in the blood coagulation mechanism, under normal conditions the plasma does not appear to contain significant amounts of lipid which are functional in coagulation. This was demonstrated by Marciniak et al. (140), who were able to infuse rather large amounts of factor X_a into mice without provoking thrombus formation. When phospholipid was added to the factor X_a, massive thrombus formation occurred, resulting in death of the animals.

IV. Termination of Coagulation

The stopping of the coagulation process is just as important as its initiation. Several mechanisms are available to prevent coagulation, once it has started, from spreading throughout the circulation: (a) Most of the clotting factors are trapped within the fibrin clot and are not free to circulate. Some of these factors, such as thrombin, are directly absorbed in the fibrin mesh. (b) The large amounts of thrombin which are produced during clotting destroy the activity of factors V and VIII, thus preventing further thrombin formation. (c) There are in plasma a variety of inhibitors of the activated coagulation factors. Perhaps the most studied of these is called antithrombin III. It results in the slow inactivation of thrombin, but more importantly, it has been shown that it has an even higher affinity for activated factor X (141, 142). Yin and Wessler (142) have shown that 1 μg of this inhibitor will inactivate 32 units of factor X_a, but only 1.2 units of thrombin. This is especially important because under optimal conditions one unit of factor X_a can lead to the formation of 50 units of thrombin. Thus, by having a high affinity for factor X_a, the antithrombin III provides a much more efficient means of inhibiting coagulation than if it were to act at a later stage. Antithrombin III can also interact with heparin to form an inhibitor which rapidly inactivates thrombin and factor X_a (141, 142). (d) During coagulation the high levels of thrombin produced may cause the generation of new inhibitory activity (143).

It seems likely that any of the activated coagulation factors which escape from the localized area of the clot would rapidly be complexed by the various plasma inhibitors and also cleared from the circulation by the liver or spleen (144). In connection with this, it should be noted that during aggregation the platelets release large amounts of glycosidases (21) so that glycoproteins in the vicinity of the thrombus would be likely to have their carbohydrate chains at least partially degraded. The observation by Morell et al. (145) that the liver rapidly clears from the circulation glycoproteins which have lost their terminal sialic acid suggests this may be an important mechanism for removal of the activated coagulation factors.

Also, since the platelet aggregate is pretty much localized at the site of the thrombus, any activated factors which escape from the region of the clot would not be very efficient in promoting coagulation at a distant site, because a phospholipid surface is necessary for their activity.

Astrup (146) has suggested that there is a continual low-level generation of thrombin and deposition of fibrin within the vascular system. This would explain the high rate of turnover of some of the coagulation factors, such as factor VIII, and more importantly would help to answer the riddle of where the first traces of thrombin, which are necessary for the modification of factor VIII, come from.

V. The Cascade Concept and Other Biological Processes

It appears very likely that the cascade mechanism and biological amplification may be occurring in other complex systems related to coagulation. For instance in plasma, those reactions which lead to plasminogen activation and eventually fibrin degradation by plasmin very likely involve a cascade effect. Indeed, there are some indications that these reactions may also involve factor XII (147).

The complement system also appears to involve a typical cascade type of mechanism [see Müller-Eberhard (148) for a recent review]. Complement refers to eleven (or more) plasma proteins which, in the presence of antigen–antibody complexes, are capable of irreversibly damaging cell membranes. Thus, the complement system plays an important role in the body's defense mechanism against infective agents. Native complement components, like plasma coagulation factors, are inactive in plasma. For physiological activity, the various components of complement are activated by specific enzymes which often appear to be proteases. Through a series of reactions, many of which involve complexes, a product is formed which leads to final membrane damage. In one of these activation pathways, for instance, a C3 proactivator convertase converts C3 proactivator to C3 activator plus a small protein fragment. The C3 activator then converts C3 to activated C3 (149, 150). Thus, some of the reactions involved in complement activation are quite analogous to blood coagulation in that inactive precursors are converted to active enzymes in a stepwise manner. Recently, it has been reported by Müller-Eberhard and co-workers (151) that the complement system may be quite closely related to blood coagulation in that rabbits with an inherited deficiency of the sixth component of complement (C6) also have a coagulation defect. Furthermore, this clotting defect was corrected *in vitro* by isolated C6.

These cascade mechanisms may also involve other important physiological reactions in plasma, such as those leading to kinin formation. Kinins are plasma substances which cause pain, increase capillary permeability, and cause smooth muscle contraction (*152–155*). The formation of kinin also involves factor XII_a, which converts a plasma precursor (kallikreinogen) into an active enzyme (kallikrein). This enzyme then releases the pharmacologically active peptides via proteolysis of kininogen (*156, 157*).

The cascade mechanism also may play an important role in the regulation of cellular reactions, particularly those involving trace components, such as hormones. For instance, in the control of glycogen degradation, there is a series of reactions which connect the hormonal stimulation of muscle with glycogen degradation [see Fischer *et al.* (*158*) for a recent review]. Thus, trace amounts of epinephrine elevate the intracellular level of cyclic AMP. Cyclic AMP then activates protein kinase, which in turn phosphorylates phosphorylase kinase. The phosphorylase kinase then converts inactive phosphorylase *b* to the active form, phosphorylase *a*, leading to glycogen degradation and the formation of glucose 1-phosphate. This series of reactions, which involves covalent modification of intermediates, differs from blood coagulation in that these reactions are reversible. Nevertheless, they are similar in that trace amounts of a hormone can lead to the formation of large amounts of glucose 1-phosphate via an amplification mechanism. It appears likely that more examples of this type involving hormones will be discovered.

REFERENCES

1. Caspary, E. A., and Kekwick, R. A., *Biochem. J.* **67**, 41 (1957).
2. Shulman, S., *J. Amer. Chem. Soc.* **75**, 5846 (1953).
3. Lanchantin, G. F., Hart, D. W., Friedman, J. A., Saavedra, N. V., and Mehl, J. W., *J. Biol. Chem.* **243**, 5479 (1968).
4. Mann, K. G., Heldebrant, C. M., and Fass, D. N., *J. Biol. Chem.* **246**, 6106 (1971).
5. Cox, A. C., and Hanahan, D. J., *Biochim. Biophys. Acta* **207**, 49 (1970).
6. Nemerson, Y., and Pitlick, F. A., *Biochemistry* **9**, 5100 (1970).
7. Esnouf, M. P., and Jobin, F., *Biochem. J.* **102**, 660 (1967).
8. Prydz, H., *Scand. J. Clin. Lab. Invest.* **17**, Suppl. 84, 78 (1965).

9. Schmer, G., Kirby, E. P., Teller, D. C., and Davie, E. W., *J. Biol. Chem.* **247**, 2512 (1972).
10. Legaz, M. E., Schmer, G., Counts, R. B., and Davie, E. W., *J. Biol. Chem.* **248**, 3946 (1973).
11. Fujikawa, K., Thompson, A. R., Legaz, M. E., Meyer, R. G., and Davie, E. W., *Biochemistry* in press (1973).
12. Østerud, B., Berre, A., Otnaess, A.-B., Bjørklid, E., and Prydz, H., *Biochemistry* **11**, 2853 (1972).
13. Jackson, C. M., and Hanahan, D. J., *Biochemistry* **7**, 4506 (1968).
14. Schiffman, S., and Lee, P., *Abstr., 3rd Cong., Int. Soc. Thromb. Haemostasis, 1972*, p. 144 (1972).
15. Schoenmakers, J. G. G., Matze, R., Haanen, C., and Zilliken, F., *Biochim. Biophys. Acta* **101**, 166 (1965).
16. Speer, R. J., Ridgway, H., and Hill, J. M., *Thromb. Diath. Haemorrh.* **14**, 1 (1965).
17. Takagi, T., and Konishi, K., *Biochim. Biophys. Acta* **271**, 363 (1972).
18. Schwartz, M. L., Pizzo, S. V., Hill, R. L., and McKee, P. A., *J. Biol. Chem.* **246**, 5851 (1971).
19. Schwartz, M. L., Pizzo, S. V., Hill, R. L., and McKee, P. A., *Abstr., 3rd Cong., Int. Soc. Thromb. Haemostasis, 1972*, p. 144 (1972).
20. Wright, I., *J. Amer. Med. Ass.* **170**, 325 (1959).
21. Day, H. J., and Holmsen, H., *Ser. Haematol.* **4**, 3 (1971).
22. Marcus, A. J., Zucker-Franklin, D., Safier, L. B., and Ullman, H. L., *J. Clin. Invest.* **45**, 14 (1966).
23. Walsh, P. N., *Brit. J. Haematol.* **22**, 237 (1972).
24. Markwardt, F., *in* "Biochemistry of Blood Platelets" (E. Kowalski and S. Niewiarowski, eds.), p. 105. Academic Press, New York, 1967.
25. Esnouf, M. P., and Macfarlane, R. G., *Advan. Enzymol. Relat. Areas Mol. Biol.* **30**, 255 (1968).
26. Davie, E. W., Hougie, C., and Lundblad, R. L., *in* "Recent Advances in Blood Coagulation" (L. Poller, ed.), p. 13. Churchill, London, 1969.
27. Hemker, H. C., Loeliger, E. A., and Veltkamp, J. J., eds., "Human Blood Coagulation. Biochemistry, Clinical Investigation and Therapy." Leiden Univ. Press, Leiden, The Netherlands, 1969.
28. Davie, E. W., and Ratnoff, O. D., *Science* **145**, 1310 (1964).
29. Macfarlane, R. G., *Nature (London)* **202**, 498 (1964).
30. Ratnoff, O. D., and Rosenblum, J. M., *Amer. J. Med.* **25**, 160 (1958).
31. Biggs, R., Sharp, A. A., Margolis, J., Hardisty, R. M., Steward, J., and Davidson, W. M., *Brit. J. Haematol.* **4**, 177 (1958).
32. Lewis, J. H., and Merchant, W. R., *J. Clin. Invest.* **37**, 911 (1958).
33. Nossel, H. L., "The Contact Phase of Blood Coagulation." Davis, Philadelphia, 1964.
34. Rosenthal, R. L., Dreskin, O. H., and Rosenthal, M., *Proc. Soc. Exp. Biol. Med.* **82**, 171 (1953).
35. Hardisty, R. M., and Margolis, J., *Brit. J. Haematol.* **5**, 203 (1959).

36. Ratnoff, O. D., Davie, E. W., and Mallett, D. L., *J. Clin. Invest.* **10,** 803 (1961).

37. Soulier, J.-P., Prou-Wartelle, O., and Menache, D., *Rev. Fr. Etud. Clin. Biol.* **3,** 263 (1958).

38. Kingdon, H. S., Davie, E. W., and Ratnoff, O. D., *Biochemistry* **3,** 166 (1964).

39. Waaler, B. A., *Scand. J. Clin. Lab. Invest.* **11,** Suppl. 37, 1 (1959).

40. Ratnoff, O. D., and Davie, E. W., *Biochemistry* **1,** 677 (1962).

41. Schiffman, S., Rapaport, S. I., and Patch, M. J., *Blood* **22,** 733 (1963).

42. Cattan, A. D., and Denson, K. W. E., *Thromb. Diath. Haemorrh.* **11,** 155 (1964).

43. Yin, E. T., and Duckert, F., *Thromb. Diath. Haemorrh.* **6,** 224 (1961).

44. Schiffman, S., Rapaport, S. I., and Patch, M. J., *Clin. Res.* **12,** 110 (1964).

45. Kingdon, H. S., *J. Biomed. Mater. Res.* **3,** 25 (1969).

46. Fujikawa, K., and Davie, E. W., unpublished observations.

47. Lundblad, R. L., and Davie, E. W., *Biochemistry* **3,** 1720 (1964).

48. Schiffman, S., Rapaport, S. I., and Chong, M. M. Y., *Proc. Soc. Exp. Biol. Med.* **123,** 736 (1966).

49. Hougie, C., Denson, K. W. E., and Biggs, R., *Thromb. Diath. Haemorrh.* **18,** 211 (1967).

50. Biggs, R., and Macfarlane, R. G., *Thromb. Diath. Haemorrh., Suppl.* **17,** 23 (1965).

51. Barton, P. G., *Nature (London)* **202,** 498 (1967).

52. Østerud, B., and Rapaport, S. I., *Biochemistry* **9,** 1854 (1970).

53. Rapaport, S. I., Schiffman, S., Patch, M. J., and Ames, S. B., *Blood* **21,** 221 (1963).

54. Macfarlane, R. G., Biggs, R., Ash, B. J., and Denson, K. W. E., *Brit. J. Haematol.* **10,** 530 (1964).

55. Rapaport, S. I., Hjort, P. F., and Patch, M. J., *Scand. J. Clin. Lab. Invest.* **17,** Suppl. 84, 88 (1965).

56. Özge-Anwar, A. H., Connell, G. E., and Mustard, J. F., *Blood* **26,** 500 (1965).

57. Caldwell, M. J., *Thromb. Diath. Haemorrh.* **17,** 256 (1967).

58. Thompson, A. R., Ph.D. Thesis, University of Washington, Seattle (1971).

59. Biggs, R., Macfarlane, R. G., Denson, K. W. E., and Ash, B. J., *Brit. J. Haematol.* **11,** 276 (1965).

60. Weiss, H. J., and Kochwa, S., *Brit. J. Haematol.* **18,** 89 (1970).

61. Østerud, B., Rapaport, S. I., Schiffman, S., and Chong, M. M. Y., *Brit. J. Haematol.* **21,** 643 (1971).

62. Niemetz, J., and Nossel, H. L., *Brit. J. Haematol.* **16,** 337 (1969).

63. Nemerson, Y., and Spaet, T. H., *Blood* **23,** 657 (1964).

64. Nemerson, Y., *J. Clin. Invest.* **47,** 72 (1968).

65. Pitlick, F. A., and Nemerson, Y., *Biochemistry* **9,** 5105 (1970).

66. Hvatum, M., Hovig, T., and Prydz, H., *Thromb. Diath. Haemorrh.* **21**, 223 (1969).
67. Hufnagel, L. A., and Riddle, J. M., *Abstr., 3rd Cong., Int. Soc. Thromb. Haemostasis, 1972*, p. 102 (1972).
68. Zeldis, S. M., Nemerson, Y., Pitlick, F. A., and Lentz, T. L., *Science* **175**, 766 (1972).
69. Pitlick, F. A., Nemerson, Y., Gottlieb, A. J., Gordon, R. G., and Williams, W. J., *Biochemistry* **10**, 2650 (1971).
70. Williams, W. J., and Norris, D. G., *J. Biol. Chem.* **241**, 1847 (1966).
71. Gladhaug, A., and Prydz, H., *Biochim. Biophys. Acta* **215**, 105 (1970).
72. Johnston, C. L., Jr., and Hjort, P. F., *J. Clin. Invest.* **40**, 743 (1961).
73. Prydz, H., *Scand. J. Clin. Lab. Invest.* **16**, 101 (1964).
74. Altman, R., and Hemker, H. C., *Thromb. Diath. Haemorrh.* **18**, 525 (1967).
75. Nemerson, Y., *Biochemistry* **5**, 601 (1966).
76. Fujikawa, K., Legaz, M. E., and Davie, E. W., *Biochemistry* **11**, 4882 (1972).
77. Esnouf, M. P., and Williams, W. J., *Biochem. J.* **84**, 62 (1962).
78. Seegers, W. H., McCoy, L. E., Reuterby, J., Sakuragawa, N., Murano, G., and Agrawal, B. B. L., *Thromb. Res.* **1**, 209 (1972).
79. Jackson, C. M., *Fed. Proc., Fed. Amer. Soc. Exp. Biol.* **31**, 241 (1972). (abstr.).
80. Fujikawa, K., Legaz, M. E., and Davie, E. W., *Biochemistry* **11**, 4892 (1972).
81. Fujikawa, K., and Davie, E. W., in preparation.
82. Titani, K., Hermodson, M. A., Fujikawa, K., Ericsson, L. H., Walsh, K. A., Neurath, H., and Davie, E. W., *Biochemistry* **11**, 4899 (1972)
83. Walsh, K. A., and Neurath, H., *Proc. Nat. Acad. Sci. U.S.* **52**, 884 (1964).
84. Magnusson, S., *in* "The Enzymes" (P. D. Boyer, ed.), 3rd ed., Vol. 3, p. 277. Academic Press, New York, 1971.
85. Hartley, B. S., *Nature (London)* **201**, 1284 (1964).
86. Robbins, K. C., Arzadon, L., Bernabe, P., and Summaria, L., *Fed. Proc., Fed. Amer. Soc. Exp. Biol.* **31**, 446 (1972) (abstr.).
87. Hartley, B. S., and Shotton, D. M., *in* "The Enzymes" (P. D. Boyer, ed.), 3rd ed., Vol. 3, p. 323. Academic Press, New York, 1971.
88. Papahadjopoulos, D., Hougie, C., and Hanahan, D. J., *Biochemistry* **3**, 264 (1964).
89. Colman, R. W., *Biochemistry* **8**, 1438 (1969).
90. Philip, G., Moran, J., and Colman, R. W., *Biochemistry* **9**, 2212 (1970).
91. Newcomb, T. F., and Hoshida, M., *Scand. J. Clin. Lab. Invest., Suppl.* **84**, 61 (1965).
92. Barton, P. G., and Hanahan, D. J., *Biochim. Biophys. Acta* **133**, 506 (1967).
93. Jobin, F., and Esnouf, M. P., *Biochem. J.* **102**, 666 (1967).

94. Schiffman, S., Theodor, I., and Rapaport, S. I., *Biochemistry* **8**, 1397 (1969).
95. Esmon, C. T., and Jackson, C. M., *Abstr., 3rd Congr., Int. Soc. Thromb. Haemostasis, 1972*, p. 88 (1972).
96. Colman, R. W., *Biochemistry* **8**, 1445 (1969).
97. Colman, R. W., Moran, J., and Philip, G., *J. Biol. Chem.* **245**, 5941 (1970).
98. Breckenridge, R. T., and Ratnoff, O. D., *J. Clin. Invest.* **44**, 302 (1965).
99. Colman, R. W., *Brit. J. Haematol.* **19**, 675 (1970).
100. Barton, P. G., Jackson, C. M., and Hanahan, D. J., *Nature (London)* **214**, 923 (1967).
101. Hemker, H. C., Esnouf, M. P., Hemker, P. W., Swart. A. C. W., and Macfarlane, R. G., *Nature (London)* **215**, 248 (1967).
102. Papahadjopoulos, D., and Hanahan, D. J., *Biochim. Biophys. Acta* **90**, 436 (1964).
103. Cole, E. R., Koppel, J. L., and Olwin, J. H., *Thromb. Diath. Haemorrh.* **14**, 431 (1965).
104. Barton, P. G., and Hanahan, D. J., *Biochim. Biophys. Acta* **187**, 319 (1969).
105. Prentice, C. R. M., and Ratnoff, O. D., *Brit. J. Haematol.* **16**, 291 (1969).
106. Prentice, C. R. M., Ratnoff, O. D., and Breckenridge, R. T., *Brit. J. Haematol.* **13**, 898 (1967).
107. Seegers, W. H., Murano, G., and McCoy, L., *Thromb. Diath. Haemorrh.* **23**, 26 (1970).
108. Magnusson, S., in "Structure-Function Relationships of Proteolytic Enzymes" (P. Desnuelle, H. Neurath, and M. Ottesen, eds.), p. 138. Academic Press, New York, 1970.
109. Magnusson, S., *Abstr., 3rd Cong., Int. Soc. Thromb. Haemostasis, 1972*, p. 3 (1972).
110. Laki, K., and Gladner, J. A., *Physiol. Rev.* **44**, 127 (1964).
111. Laki, K., in "Fibrinogen" (K. Laki, ed.), p. 1. Dekker, New York, 1968.
112. Blomback, B., and Doolittle, R. F., *Acta Chem. Scand.* **17**, 1819 (1963).
113. Blomback, B., Blomback, M., and Grondahl, N. J., *Acta Chem. Scand.* **19**, 1789 (1965).
114. Blomback, B., Blomback, M., Grondahl, N. J., and Holmberg, E., *Ark. Kemi* **25**, 411 (1966).
115. Doolittle, R. F., Schubert, D., and Schwartz, S. A., *Arch. Biochem. Biophys.* **118**, 456 (1967).
116. Hann, C. S., *Biochim. Biophys. Acta* **124**, 398 (1966).
117. Osbahr, A. J., Colman, R. W., and Patsy, S. M., *Biochem. Biophys. Res. Commun.* **25**, 309 (1966).
118. Herzig, R. H., Ratnoff, O. D., and Shainoff, J. R., *J. Lab. Clin. Med.* **76**, 451 (1970).
119. Lorand, L., and Konishi, K., *Fed. Proc., Fed. Amer. Soc. Exp. Biol.* **21**, 62 (1962).

120. Lorand, L., Konishi, K., and Jacobsen, A., *Nature (London)* **194**, 1148 (1962).

121. Konishi, K., and Lorand, L., *Biochim. Biophys. Acta* **121**, 177 (1966).

122. Lorand, L., *Thromb. Diath. Haemorrh., Suppl.* **13**, 45 (1965).

123. Lorand, L., and Ong, H. H., *Biochem. Biophys. Res. Commun.* **23**, 188 (1966).

124. Lorand, J., Ong, H. H., Lipinski, B., Rule, N. G., Downey, J., and Jacobsen, A., *Biochem. Biophys. Res. Commun.* **25**, 629 (1966).

125. Lorand, L., Urayama, T., and Lorand, L., *Biochem. Biophys. Res. Commun.* **23**, 828 (1966).

126. Loewy, A. G., Matacic, S., and Darnell, J. H., *Arch. Biochem. Biophys.* **113**, 435 (1966).

127. Tyler, H. M., and Laki, K., *Biochem. Biophys. Res. Commun.* **24**, 506 (1966).

128. Fuller, G. M., and Doolittle, R. F., *Biochem. Biophys. Res. Commun.* **25**, 694 (1966).

129. Doolittle, R. F., Chen, R., and Lau, F., *Biochem. Biophys. Res. Commun.* **44**, 94 (1971).

130. Pisano, J. J., Finlayson, J. S.. Peyton, M. P., and Nagai, Y., *Proc. Nat. Acad. Sci. U.S.* **68**, 770 (1971).

131. Ball, A. P., Hill, R. L., and McKee, P. A., *Abstr., 3rd Congr., Int. Soc. Thromb. Haemostasis, 1972,* p. 62 (1972).

132. Gormsen, J., Fletcher, A. P., Alkjaersig, N., and Sherry, S., *Arch. Biochem. Biophys.* **120**, 654 (1967).

133. Mattiasson, B., and Mosbach, K., *Biochim. Biophys. Acta* **235**, 253 (1971).

134. Smith, R., and Tanford, C., *J. Mol. Biol.* **67**, 75 (1972).

135. Wallach, D. F. H., Maurice, P. A., Steele, B. B., and Surgenor, D. M., *J. Biol. Chem.* **234**, 2829 (1959).

136. Bangham, A. D., *Nature (London)* **192**, 1197 (1961).

137. Papahadjopoulos, D., Hougie, C., and Hanahan, D. J. *Proc. Soc. Exp. Biol. Med.* **111**, 412 (1962).

138. Daemen, F. J. M., and van Deenen, L. L., *Biochem. J.* **88**, 32P (1963).

139. Hemker, H. C., Kahn, M. J. P., AND Devilee, P. P., *Thromb. Diath. Haemorrh.* **24**, 214 (1970).

140. Marciniak, E., Rodriguez-Erdman, F., and Seegers, W. H., *Science* **137**, 421 (1962).

141. Biggs, R., Denson, K. W. E., Adman, N., Borrett, R., and Hadden, M., *Brit. J. Haematol.* **19**, 283 (1970).

142. Yin, E. T., Wessler, S., and Stoll, P. J., *J. Biol. Chem.* **246**, 3694, 3703, and 3712 (1971).

143. Marciniak, E., *J. Lab. Clin. Med.* **79**, 924 (1972).

144. Spaet, T. H., *Blood* **28**, 112 (1966).

145. Morell, A. G., Gregoriadis, G., Scheinberg, I. H., Hickman, J., and Ashwell, G., *J. Biol. Chem.* **246**, 1461 (1971).

146. Astrup, T., *Thromb. Diath. Haemorrh.* **2**, 347 (1958).

147. Iatrides, S. G., and Ferguson, J. H., *J. Clin. Invest.* **41**, 1277 (1962).

148. Müller-Eberhard, H. J., *Annu. Rev. Biochem.* **38,** 389 (1969).

149. Götze, O., and Müller-Eberhard, H. J., *J. Exp. Med.* **134,** 90 (1971).

150. Müller-Eberhard, H. J., and Götze, O., *J. Exp. Med.* **135,** 1003 (1972).

151. Müller-Eberhard, H. J., Zimmerman, T. S., and Taylor, F. B., *Abstr., 3rd Congr. Int. Soc. Thromb. Haemostasis, 1972,* p. 4 (1972).

152. Margolis, J., *Ann. N.Y. Acad. Sci.* **104,** 133 (1963).

153. Margolis, J., *J. Physiol. (London)* **151,** 238 (1960).

154. Webster, M. E., and Ratnoff, O. D., *Nature (London)* **192,** 180 (1961).

155. Ratnoff, O. D., and Miles, A. A., *Brit. J. Exp. Pathol.* **45,** 328 (1964).

156. Colman, R. W., Mattler, L., and Sherry, S., *J. Clin. Invest.* **48,** 11 (1969).

157. Colman, R. W., Mattler, L., and Sherry, S., *J. Clin. Invest.* **48,** 23 (1969).

158. Fischer, E. H., Heilmeyer, L. M. G., and Haschke, R. H., *Curr. Top. Cell Regul.* **4,** 211 (1971).

Enzymatic ADP-Ribosylation of Proteins and Regulation of Cellular Activity

TASUKU HONJO*

Department of Embryology, Carnegie
Institution of Washington, Baltimore, Maryland

OSAMU HAYAISHI

Department of Medical Chemistry
Kyoto University Faculty of Medicine, Kyoto, Japan

I. Introduction

Nicotinamide-adenine dinucleotide (NAD) has firmly been established as a coenzyme in biological oxidation. This coenzyme, however, has been recently found to participate in several other important reactions. Deoxyribonucleic acid (DNA) ligase of *Escherichia*

* *Present address:* Laboratory of Molecular Genetics, National Institute of
Child Health and Human Development, National Institute of Health,
Bethesda, Maryland 20014.

coli requires NAD as a cofactor to synthesize phosphodiester bonds between the 5'-phosphoryl and 3'-hydroxyl termini of "nicked" double stranded DNA (*85, 122*). NAD serves as an energy donor, being cleaved to adenosine 5'-phosphate (AMP) and micotinamide mononucleotide (NMN). In the first step of this reaction, the AMP portion of NAD is transferred to the amino group of a lysine residue in the enzyme molecule with a simultaneous release of NMN (*42*). In another class of reactions utilizing NAD, the adenosine 5'-diphosphate (ADP)-ribose portion is transferred to an acceptor protein molecule (Fig. 1). Diphtheria toxin-catalyzed ADP-ribosylation of elongation factor 2 (EF 2) is an example of this class of reaction (*50–52*). ADP-ribosylation of EF 2 results in the inactivation of the enzyme, which is now believed to be the molecular basis of the toxic effect of diphtheria toxin. The unique modification of EF 2 has provided a model for a possible regulatory mechanism in polypeptide synthesis, although similar reactions have not yet

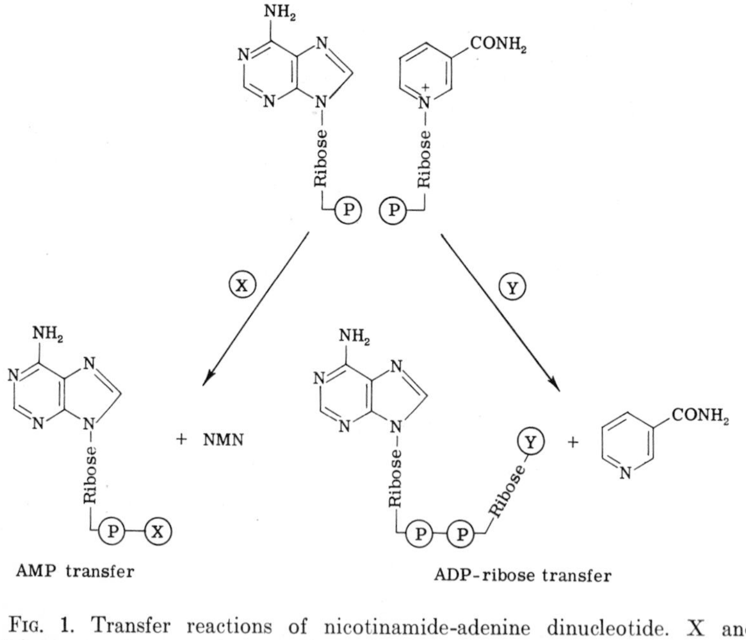

FIG. 1. Transfer reactions of nicotinamide-adenine dinucleotide. X and Y represent the acceptor protein molecules described in the text. NMN, nicotinamide mononucleotide. Reproduced from Nishizuka *et al.* (*82*) by permission of the Cold Spring Harbor Laboratory.

been discovered in normal cells. The discovery of the catalytic activity of the toxin has prompted extensive investigations of the structure–activity relationship of the toxin molecule. The other known example of an ADP-ribosylation reaction is that of the nuclear proteins (*13, 25, 80, 93, 109*), which is catalyzed by an enzyme associated with mammalian chromatin.

This chapter deals primarily with various aspects of ADP-ribosylation of EF 2. The reaction of nuclear proteins will also be described briefly, although its biological function is still obscure. ADP-ribosylation of EF 2 was previously reviewed in Japanese (*49*), and ADP-ribosylation of nuclear proteins was also covered by a recent review (*108*).

II. ADP-Ribosylation of EF 2 by Diphtheria Toxin

A. Function of EF 2 in Polypeptide Synthesis

Elongation factors 1 and 2 are two complementary enzymes which participate in the elongation of polypeptide chains in eukaryotic cells. Ribosomes are believed to have at least two sites at which tRNA molecules can bind with proper anticodon–codon pairing. After the formation of a peptide bond between a nascent polypeptide chain and an amino acid, the peptidyl-tRNA is now located on the so-called "acceptor" site. In order to elongate the polypeptide chain, peptidyl-tRNA must be translocated to the "donor" site, leaving the acceptor site available to the next aminoacyl-tRNA directed by mRNA. EF 1 catalyzes binding of aminoacyl-tRNA to the acceptor site on ribosomes. EF 2 plays a role in the translocation of peptidyl-tRNA coupled with hydrolysis of guanosine 5′-triphosphate (GTP) to guanosine 5′-diphosphate (GDP) and P_i. EF 2 is also known as aminoacyltransferase II, transfer factor II or T_2. EF 2 corresponds to G factor or EF G in bacterial systems. For a more detailed description of these events, the reader is referred to a recent review (*64*).

B. Historical Sketch

Diphtheria toxin is an exotoxin produced by *Corynebacterium diphtheriae* lysogenic for phage β. It is a simple protein consisting of a single polypeptide chain of approximately 62,000 daltons. The toxin, which has been purified and crystallized from the culture medium of a toxin-producing strain of *C. diphtheriae*, is highly lethal to humans, horses, rabbits, or guinea pigs. The minimal lethal dose for humans is reported to be 2.5 μg. Since the toxin was discovered in 1888 by Roux and Yersin (*95*), a number of attempts have been made to reveal the mechanism of the lethal effect of diphtheria toxin on animals. However, it was not until 1959 that the biochemical analysis of its mode of action began. This was made possible by the report of Strauss and Hendee (*107*) that the toxin inhibits protein synthesis in HeLa cells. Diphtheria toxin inhibits the incorporation of amino acids into the acid-insoluble fraction under conditions where there is no inhibition of RNA synthesis, glycolysis, or oxygen uptake, nor any change in nucleotide metabolism (*61, 107*). The toxin has also been shown to inhibit polypeptide synthesis in cell-free systems (*59*). In 1964, Collier and Pappenheimer found NAD to be absolutely required for the inhibition of polypeptide synthesis by diphtheria toxin (*18*).

Subsequently, Collier showed that the toxin specifically inactivates EF 2 in a cell-free system from rabbit reticulocytes (*15*). He separated supernatant enzymes into two fractions, EF 1 and 2, and then treated each one of them with the toxin and NAD. EF 2 was inactivated by this treatment, whereas EF 1 remained active. The inactivation of EF 2 was dependent on both toxin and NAD. The toxin did not inhibit the binding of phenylalanyl-tRNA to ribosomes in the presence of poly(U), again indicating that EF 1 is not affected by the toxin. The polysome pattern of the toxin-treated cell-free system was analyzed on sucrose density gradients. No breakdown of polysomes was observed compared with the normal system. The polysomes thus isolated from the toxin-treated system were as active as the normal polysomes. Synthesis of aminoacyl-tRNA was not affected by the toxin (*18*). It is clear, therefore, that toxin inactivates EF 2 specifically without affecting EF 1, ribosomes, or amino acid-activating enzymes.

Pappenheimer and his colleagues confirmed and extended this conclusion using the cell-free system from HeLa Cells (*39–41*). Goor, Pappenheimer, and Ames (*41*) studied the effect of varying toxin concentrations on amino acid incorporation at constant NAD concentrations and also the effect of various NAD concentrations at constant toxin concentrations. The following observations were reported. (a) There is an inverse relationship between the concentrations of the toxin and NAD. The higher the NAD concentration, the less toxin is required for the inhibition of protein synthesis. (b) At low concentrations of toxin (4×10^{-9} M) maximal inhibition of polypeptide synthesis cannot be reached even at a high NAD concentration (2×10^{-5} M). (c) The inhibition of polypeptide synthesis is partially reversible by the addition of a 1000-fold excess amount of nicotinamide over NAD. Based on these observations, especially (a) and (b), they proposed that the toxin interacts with NAD and EF 2 to form a ternary complex which is catalytically inactive. They assumed Eqs. (1) and (2) for the formation of the ternary complex and calculated the percent inhibition of protein synthesis as Eq. (3) where K_1 and K_2 are association constants of Eqs. (1) and (2), respectively. The titration

$$\text{Toxin} + \text{NAD} \rightleftarrows \text{toxin-NAD} \qquad (1)$$

$$\text{Toxin-NAD} + \text{EF 2} \rightleftarrows \text{EF 2-toxin-NAD} \qquad (2)$$

$$\% \text{ inhibition} = \frac{100 \text{ (toxin) (NAD)} K_1 K_2}{1 + (\text{NAD})K_1 + (\text{toxin}) (\text{NAD})K_1 K_2} \qquad (3)$$

curves of NAD or the toxin were best fitted for the curves predicted from Eq. (3) using values of 2.4×10^3 and 1.4×10^{10} liter/mole for K_1 and K_2, respectively.

The ternary complex theory, however, cannot explain the reactivation of toxin-treated cell extracts by nicotinamide. Furthermore, none of these observations described above exclude the possibility that the toxin might function catalytically, because they did not consider the possibility that the inactivation of EF 2 is not instantaneous.

C. ADP-Ribosylation of EF 2

Our group in Kyoto had long been engaged in studies on the metabolism of NAD and became interested in the role of NAD in the inactivation of EF 2 by diphtheria toxin. The fact that nicotinamide can restore the inactivated EF 2 activity suggested to us that NAD does not function as a cofactor of a dehydrogenase reaction. In order to explore the fate of NAD during inactivation of EF 2, we prepared NAD labeled with ^{14}C in the adenine moiety. When NAD-(adenine)-^{14}C was incubated with the toxin and partially purified EF 2 from rat liver, the radioactivity was made acid insoluble (50,

TABLE I
INCORPORATION OF RADIOACTIVITY INTO ACID-INSOLUBLE
FRACTION FROM NAD-(ADENINE)-^{14}C^a

System	Acid-insoluble radioactivity (cpm)
EF 2 alone	17
Diphtheria toxin alone	0
EF 2 + diphtheria toxin	1973
EF 1 + diphtheria toxin	3
EF 2 + diphtheria toxin + nicotinamide	256
EF 2 + diphtheria toxoid	28
EF 2 + diphtheria toxin + diphtheria antitoxin	4

^{a} The complete reaction mixture (0.1 ml) contained 10 μmoles
of Tris·HCl buffer, pH 7.6, 196 pmoles of NAD-(adenine)-^{14}C
(38.8 cpm/pmole), 200 μg of EF 2, and 2 μg (1 flocculation unit)
of diphtheria toxin. Incubation was carried out for 10 minutes at
37°C. Reaction was stopped by the addition of 5% trichloroacetic acid and acid-insoluble radioactivity was determined.
Where indicated, 268 μg (2 international units) of antitoxin, 31 μg
(10 flocculation units) of diphtheria toxoid, 10 μmoles of nicotinamide, and 60 μg of EF 1 were added. Reproduced from Honjo
et al. (50) by permission of the Cold Spring Harbor Laboratory.

51). The results are shown in Table I. The incorporation of radioactivity into acid-insoluble material was dependent on both toxin
and EF 2. EF 1 did not replace EF 2. Nicotinamide inhibited the
incorporation of radioactivity from NAD-(adenine)-^{14}C into the
acid-insoluble fraction. Diphtheria toxoid, which is immunogenic
but not toxic, was inactive in this reaction and diphtheria antitoxin
counteracted the activity of toxin completely.

We then synthesized NAD preparations labeled in different
moieties in order to determine which portion of the NAD molecule
is incorporated into the acid-insoluble product (50, 51). Table II
shows that when NAD labeled either with ^{14}C in adenine, with ^{3}H
in adenosine, with ^{32}P, or with ^{14}C in ribose of NMN, was used
as substrate, essentially the same amount of radioactive material
was incorporated into the acid-insoluble fraction. However, when
NAD labeled with ^{14}C in nicotinamide was used as substrate, no

TABLE II
INCORPORATION OF LABELED NAD PREPARATIONS
INTO ACID-INSOLUBLE FRACTION[a]

NAD employed	Incorporation into acid-insoluble fraction (pmoles)
NAD-(adenine)-^{14}C	51.2
NAD-(adenosine)-^{3}H	50.6
NAD-(both phosphates)-^{32}P	50.0
NAD-(ribose in NMN)-^{14}C	50.0
NAD-(nicotinamide)-^{14}C	0.3

[a] The reaction was carried out as described in Table I, except that 400 pmoles of various radioactive NAD's were employed. Reproduced from Honjo et al. (50) by permission of the Cold Spring Harbor Laboratory.

incorporation of radioactivity was observed in acid-insoluble material. Therefore, only the ADP-ribose portion of NAD is incorporated into the acid-insoluble fraction. A stoichiometric amount of nicotinamide is released simultaneously into the acid-soluble fraction (52).

The reaction product prepared from NAD-(ribose in NMN)-^{14}C is fairly stable (50–52). It resists heat treatment at 95°C for 6 minutes in 0.1 N HCl or 0.1 N NaOH. However, when it is heated at 95° in 1 N HCl for 10 and 30 minutes, the conversion of the product to an acid-soluble form is 20 and 50%, respectively. Heating at 95°C in 1 N NaOH for 6 minutes renders 43% of the product acid-insoluble. The reaction product is not dialyzable but is sensitive to digestion by various proteases, such as trypsin, α-chymotrypsin, papain, bromelin, and Pronase. It resists digestion with pancreatic DNase, T$_1$ RNase, pancreatic RNase, α-amylase, and Neurospora NADase.

When the reaction product prepared using NAD-(adenine)-^{14}C as substrate is digested with snake venom phosphodiesterase, it is quantitatively converted to an acid-soluble form (52). The acid-soluble compound is isolated and identified as 5'-AMP. 2'-(5''-Phosphoribosyl)-5′-AMP, which is obtained upon the similar treatment of poly ADP-ribose (see Section III,A), is not detected at all. The results indicate that ADP-ribose is bound to protein as a monomer without further elongation of the ADP-ribose chain. Furthermore, ADP-ribose seems to be attached to protein through ribose portion in NMN, since 5′-AMP is released from the reaction product by phosphodiesterase digestion.

When the concentration of EF 2 is kept constant and the amount of diphtheria toxin is varied from 0.4 to 2 μg, the initial rate of incorporation increases, but the maximal amount of ADP-ribose incorporated remains essentially constant. On the other hand, when the toxin concentration is kept constant and the amount of EF 2 is changed, the rate as well as the extent of incorporation of ADP-ribose appears to be dependent on the amount of EF 2 (52). The results suggest that toxin plays a catalytic function, ADP-ribose being bound to EF 2.

In order to clearly demonstrate that ADP-ribose is attached to EF 2 but not to the toxin, the ADP-ribosylated protein was prepared in a large-scale incubation mixture containing EF 2, toxin and NAD-(adenosine)-[14]C (50, 51). The reaction product was chromatographed on a hydroxyapatitie column. As shown in Fig. 2a, acid-insoluble radioactivity was eluted as a single peak separated from that of the toxin. Similar results were obtained by the experiments with ^{125}I-labeled diphtheria toxin and NAD-(both phosphates)-[32]P (52). The ADP-ribosylated product thus isolated had essentially no EF 2 activity as measured by polyphenylalanine synthesis. However, when the product was incubated for 60 minutes at 37°C in the presence of both toxin and nicotinamide, the acid-insoluble radioactivity disappeared with a concomitant restoration of the EF 2 activity. Similar treatment with nicotinamide or toxin alone had no effect either on the acid-insoluble radioactivity or on the EF 2 activity. The acid-soluble product of the reverse reaction was isolated by chromatography on a Dowex 1 column and identified as NAD by paper chromatography and by paper electrophoresis. The results indicate that the ADP-ribosylation reaction is reversible.

Direct evidence that ADP-ribose is attached to EF 2 comes from experiments designed to see whether ADP-ribosylated protein coin-

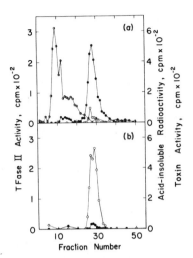

FIG. 2. Isolation of ADP-ribosyl-EF 2 by hydroxyapatite column chromatography. (a) The reaction mixture contained 7.2 nmoles of NAD-(adenosine)-¹⁴C (118 cpm/pmole), 20 μg of a crystalline preparation of diphtheria toxin, 40 mg of EF 2 (TFase II), 0.8 μmole of DTT, and 50 μmoles of Tris·HCl buffer, pH 7.6, in a total volume of 1.2 ml. After incubation for 60 minutes at 37°C, the entire mixture was chromatographed of a hydroxyapatite column. Elution was carried out with the application of a linear concentration gradient from 0.001 M to 0.2 M potassium phosphate buffer, pH 7.0, containing 1 mM 2-mercaptoethanol. The acid-insoluble radioactivity and the activity of polyphenylalanine synthesis were determined. An aliquot of each fraction was also assayed for toxin content by measuring the ability of catalyze the incorporation of NAD-(adenosine)-¹⁴C into acid-insoluble material. (b) An aliquot (50 μl) of each fraction was incubated for 60 minutes at 37°C in a reaction mixture containing 20 μg of diphtheria toxin and 6.25 μmoles of nicotinamide in a total volume of 0.125 ml. Then, the acid-insoluble radioactivity of ADP-ribose and the activity of polyphenylalanine synthesis were again determined. ●——● Acid-insoluble radioactivity of ADP-ribose; ○——○, polyphenylalanine synthesis; ◑——◑, toxin activity. Reproduced from Honjo et al. (51) by permission of the American Society of Biological Chemists, Inc.

cides with the EF 2 activity restored by the reverse reaction upon chromatography on an hydroxyapatite column. When an aliquot of each fraction of hydroxyapatite column chromatography (Fig. 2a) was incubated with the toxin and nicotinamide to carry out the reverse reaction, the acid-insoluble radioactivity disappeared with a concomitant appearance of the EF 2 activity in the corresponding fractions as shown in Fig. 2b.

Available evidence strongly indicates that diphtheria toxin catalyzes the transfer of ADP-ribose to EF 2, resulting in a concurrent inactivation of this enzyme, and furthermore, that treatment of ADP-ribosyl-EF 2 with both toxin and nicotinamide restores the EF 2 activity with a simultaneous release of NAD into the acid-soluble fraction. The scheme of the reaction is shown below.

$$EF\ 2 + NAD \underset{\longleftarrow}{\overset{toxin}{\longrightarrow}} ADP\text{-ribosyl-EF } 2 + nicotinamide + H^+ \quad (4)$$

Gill et al. (36) observed that diphtheria toxin releases nicotinamide from NAD in the presence of EF 2 and reached a conclusion similar to Eq. (4). ADP-ribosylation of EF 2 has since been confirmed and extended by several investigators (16, 38).

Subsequently, EF 2 was purified from rat liver to a homogeneous protein by ammonium sulfate precipitation, DEAE-Sephadex, hydroxyapatite, and electrofocusing columns as shown in Table III (52). Specific activity was enriched more than 700-fold, with an overall yield of about 25% starting from the ammonium sulfate fraction. The ratios of specific activity of EF 2 to that of the ADP-ribose acceptor in these fractions were essentially identical throughout the purification steps, indicating that EF 2 is the sole protein species to be ADP-ribosylated in the supernatant fraction from rat

TABLE III
SUMMARY OF PURIFICATION OF EF 2[a]

Fraction	Protein (mg)	EF 2 activity (units)	Specific activity (units/μg protein)
Ammonium sulfate fraction	11,900	900×10^3	0.075
DEAE-Sephadex eluate	605	575×10^3	0.95
Hydroxyapatite eluate	152	460×10^3	3.25
Electrofocusing fraction	3.86	225×10^3	58

[a] Reproduced from Honjo et al. (52) by permission of the American Society of Biological Chemists, Inc.

liver. The final preparation thus obtained was homogeneous upon polyacrylamide gel electrophoresis. One microgram of the most highly purified preparation of EF 2 is capable of accepting 5.2 and 16.5 pmoles of ADP-ribose based on the colorimetric and spectrophotometric determinations of protein concentrations, respectively. The molecular weight of EF 2 is estimated to be approximately 60,000–70,000 by its elution profile from a Sephadex G-100 column, which coincides well with the value (65,000) reported by Galasinski and Moldave (30). It follows, therefore, that approximately 1 mole of ADP-ribose binds to 1 mole of EF 2. This number is also supported by the report of Maxwell and her co-workers, who showed that EF 2 consists of a single polypeptide chain (20, 89). These investigators also reported the molecular weight of EF 2 to be approximately 100,000.

Several lines of evidence indicate that ADP-ribosylation of EF 2 is quantitatively proportional to and closely associated with the inactivation of the enzyme (50–52). The time course of incorporation of the ADP-ribose moiety of NAD and inactivation of EF 2 is shown in Fig. 3. It can be seen that the incorporation of radioactive ADP-ribose is exactly proportional to the degree of inactivation of the EF 2 activity. Essentially similar results were obtained when the reaction was run in forward or reverse direction in the presence of various concentrations of nicotinamide (52).

Several investigators (24, 62, 74) have reported that ADP-ribosylated EF 2 is unable to interact with ribosomes. ADP-ribosylated EF 2 does not compete with the active form of the enzyme in polypeptide synthesis (62). Everse et al. (24) showed that ribosomal pellets isolated from incubation mixtures containing ADP-ribosylated EF 2 and GTP are unable to support protein synthesis by themselves, but are active provided that the active form of EF 2 is added. They claimed that newly added EF 2 is able to support protein synthesis because the functional sites on ribosomes remain free of ADP-ribosylated EF 2 after preincubation. However, they did not exclude the possibility that the active form of EF 2 can replace ADP-ribosylated EF 2 bound to ribosomes. More directly, Montanaro et al. (74) showed by centrifugation through sucrose density gradient that purified ADP-ribosylated EF 2 is unable to interact with ribosomes. Contrary to these findings, ADP-

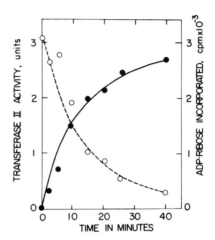

Fig. 3. Time course for the incorporation of the ADP-ribose portion of NAD into the acid-insoluble fraction and for the inactivation of EF 2 (transferase II). Each reaction mixture (0.1 ml) contained 100 μg of EF 2, 0.2 μg (0.1 flocculation unit) of a crystalline preparation of diphtheria toxin, 216 pmoles of NAD-(adenosine)-^{14}C (118 cpm/pmole), 20 nmoles of dithiothreitol, and 10 μmoles of Tris·HCl buffer, pH 7.6. Incubation was carried out at 37°C. At specified intervals the EF 2 activity and the incorporation of ADP-ribose were assayed. The incorporation of ADP-ribose reached almost maximum at 40 minutes. To assay EF 2, an aliquot (2 μl) of each reaction mixture was added to another mixture (188 μl) containing 1 international unit of diphtheria antitoxin to stop the additional incorporation of ADP-ribose, 10 μmoles of Tris·HCl buffer, pH 7.6, 1.6 μmoles of MgCl$_2$, 16 μmoles of NH$_4$Cl, 3.2 μmoles of 2-mercaptoethanol, 30 μg of ribosomes, 40 μg of poly(U), 0.1 μmole of GTP, 30 μg of ^{14}C-phenylalanyl-tRNA (3000 cpm), and 60 μg (an excess amount) of EF 1. After incubation for 15 minutes at 37°C, hot 5% trichloroacetic acid-insoluble radioactivity was determined. Under these conditions, the radioactive ADP-ribose attached to protein was released into the acid-soluble fraction, and did not interfere with the assay of polypeptide synthesis. ●——●, ADP-ribose incorporation into acid-insoluble fraction; ○---○, the activity of polyphenylalanine synthesis. Reproduced from Honjo et al. (50) by permission of the Cold Spring Harbor Laboratory.

ribosylated EF 2 is shown to bind to ribosomes on an Agarose column (49). Further work appears necessary to clearly resolve this question. Raeburn et al. (89) found that both amino acid incorporation and ribosome-dependent GTPase is considerably less stable than the ADP-ribose accepting activity of EF 2, which together with the studies with antibiotic inhibitors suggests that the site of ADP-ribose attachment might be distinct from the active site of the enzyme.

D. Properties of the ADP-Ribosylation Reaction

The ADP-ribosylation and reverse reactions are optimal at pH 8.5 and 5.2, respectively (*50*). The Michaelis constants for NAD and nicotinamide are 5 μM (pH 7.4) and 0.5 mM (pH 5.2), respectively (*52*). The ADP-ribosylation reaction, which is specific for NAD (*52*), does not accept NADP, NMN, and NADH as substrates. The thionicotinamide analog of NAD, which can replace NAD in many dehydrogenase reactions, is as active as NAD. In the reverse reaction, thionicotinamide is also capable of substituting for nicotinamide (*52*). Acetylpyridine is 10% as active as nicotinamide, while nicotinate, NMN, AMP, and ADP-ribose are totally inactive. These results are consistent with the report by Goor and Pappenheimer (*40*) that thionicotinamide and acetylpyridine analogs of NAD are active in inhibition of polypeptide synthesis by the toxin. EF 2 from yeast (*41, 44*) and wheat (*65*) as well as various animal cells appears to be ADP-ribosylated by the toxin. EF G in bacterial systems is not ADP-ribosylated. No protein other than EF 2 of eukaryotic cells has been found capable of accepting ADP-ribose from NAD in the presence of diphtheria toxin (*52*). Thus, when a crude NAD-free extract from eukaryotic cells is incubated with [32]P- or (adenine)-^{14}C-labeled NAD together with excess toxin, the acid-precipitable radioactivity will provide a measure of its EF 2 content (*52*). In fact, this method provides a useful quick assay procedure for EF 2 in place of the troublesome protein-synthesizing assay system (*33*).

Various types of inhibitors of the reaction are known. Diphtheria antitoxin inhibits the reaction completely at a concentration equivalent to that of the toxin (*51, 52*), whereas GTP (*105*), ribosomes (*38, 42*), and ribosomal RNA (*113, 114*) retard the rate of the reaction. Two groups of competitive inhibitors are known (*38*). The inhibitors in the first group, which are competitive with respect to both NAD and EF 2, includes GTP (K_i = 4.9 mM), GDP, GMP, 5'-guanylylmethylene diphosphate (GMPPCP), inosine 5'-triphosphate (ITP), NMN (K_i 3.3 mM), and nicotinamide (K_i = 0.21

mM). GTP inhibits both forward and backward reactions, whereas NMN inhibits only the forward reaction. The inhibitors in the second group are competitive with respect to NAD and noncompetitive with respect to EF 2. This group includes adenine ($K_i = 36$ μM), adenosine, AMP, ADP, ATP, and NADH ($K_i = 36$ μM). The K_i value for adenine was also reported to be 38.5 μM from equilibrium dialysis (104). NADH is shown to have a higher affinity to toxin than NAD by measuring the change of fluorescence (73). NADH inhibits both forward and reverse reactions whereas adenine inhibits only the forward reaction. Several antibiotics also inhibit the reaction (89). Fusidic acid (1 mM), emetin (10 mM), and cycloheximide (10 mM) inhibit the ADP-ribosylation reaction to the extent of 35, 25, and 30%, respectively. Neomycin (0.1 mM), sparsomycin (10 μM), and gourgerotin (1 mM) do not inhibit the reaction at all. DTT activates the catalytic activity of the toxin (16) due to the reduction of a disulfide bond of the toxin molecule. The relationship between the structure and activity of the toxin is described in Section II, H.

The extent of inhibition of protein synthesis by the toxin usually falls between 70 and 90% and rarely reaches 100%. Goor and Pappenheimer (39) suggested that the toxin might be unable to inactivate EF 2 associated with ribosomes thereby explaining the residual activity. Ribosomal RNA (113, 114) as well as ribosomes (38, 91) inhibit the initial rate of ADP-ribosylation of EF 2. Gill et al. (35) have shown that EF 2 interacts with ribosomes to form a complex in vitro and that EF 2 bound to ribosomes is no longer a substrate for the ADP-ribosylation reaction. Experiments in vivo have also been consistent with this idea (35, 103).

E. Reaction Mechanism of ADP-Ribosylation of EF 2

Since the toxin catalyzes the release of nicotinamide with the concomitant transfer of ADP-ribose to EF 2, the toxin may be considered as a type of NADase. Indeed, fragment A, which is a catalytically active polypeptide of the toxin (see Section II, H), has NAD glycohydrolase activity even in the absence of EF 2, although the activity is orders of magnitude less than the ADP-ribosylating activity (57). Two kinds of NADase have been described. Microbial

NADases, such as *Neurospora* NADase (*58*), do not catalyze trans-glycosidation but simply hydrolyze NAD to nicotinamide and ADP-ribose. Mammalian NADases, in contrast, catalyze both hydrolysis of NAD and transfer of the ADP-ribose moiety to various basic acceptors, such as pyridine derivatives (*121*). From this point of view, diphtheria toxin, though coded by the β phase genome (*115*), is similar to mammalian NADases. In the case of the latter enzymes, the first step of the reaction gives rise to an intermediate, i.e., ADP-ribosyl-enzyme, followed by transfer of ADP-ribose to various acceptors, or alternatively followed by hydrolysis to yield free ADP-ribose and the enzyme. A similar mechanism in ADP-ribosylation of EF 2 can be expressed by two partial reactions, Eqs. (5) and (6).

$$\text{Toxin} + \text{NAD} \rightleftarrows \text{ADP-ribosyl-toxin} + \text{nicotinamide} + \text{H}^+ \quad (5)$$

$$\text{ADP-ribosyl-toxin} + \text{EF 2} \rightleftarrows \text{ADP-ribosyl-EF 2} + \text{toxin} \quad (6)$$

If this be the case, the toxin, in the absence of EF 2, should catalyze two isotope exchange reactions (a) between nicotinamide-^{14}C and NAD, and (b) between ADP-^{3}H-ribosyl-EF 2 and NAD. Experimental results, however, show that the toxin is incapable of catalyzing the first exchange reaction unless EF 2 is present (*38, 52*). A very slight but significant amount of the second exchange reaction was detected even in the absence of EF 2 (*38*). The results, however, should be reexamined because the preparation of ADP-ribosylated EF 2 might have been contaminated with unreacted EF 2, which has since been shown to be separable by the electrofocusing technique (*90*). Goor and Maxwell (*38*) made kinetic studies on this reaction. The initial rate was measured at a series of different concentrations of one of the two substrates while the other substrate concentration was held constant. Lineweaver-Burk plots showed that the lines intersected on the abscissa. The mechanism defined by Eqs. (5) and (6) is a Ping-Pong mechanism and postulates parallel lines. Intersection of the lines on the abscissa is more consistent with an ordered than with a Ping-Pong mechanism. Thus, available evidence favors the ordered mechanism proposed by Goor and Maxwell (*38*) and Honjo *et al.* (*52*). This mechanism predicts

an unstable ternary complex consisting of toxin, EF 2 and NAD
as shown in Eqs. (7) and (8).

$$\text{Toxin} + \text{NAD} \rightleftarrows \text{toxin-NAD} \tag{7}$$

$$\text{Toxin-NAD} + \text{EF 2} \rightleftarrows (\text{EF 2–toxin–NAD}) \rightleftarrows$$
$$\text{ADP-ribosyl–EF 2} + \text{nicotinamide} + \text{H}^+ + \text{toxin} \tag{8}$$

NAD has been shown to bind to the toxin (*40, 104*). Fragment A
has also been shown by nuclear magnetic resonance studies (*46*)
to interact with NAD. The isolation of an EF 2–toxin–NAD complex
has been reported (*24*). The formation of this complex, however,
is probably due to catalytically inactive species of the toxin, i.e.,
intact or nicked toxin, which will be described in Section II, H.
No evidence has been obtained that the complex is an intermediate
in the ADP-ribosylation reaction. Recently, Kandel and Collier
(*57*) reported that the catalytically active fragment A of the toxin
has a very weak NAD glycohydrolase activity. This NADase activity seems to be different from the bacterial NADase associated with
some toxin preparations (*52*), because the NADase activity is found
only in the fragment A fraction and not in the intact or nicked
toxin. Could this be the first step in the ADP-ribosylation of EF
2? If so, an ordered mechanism (Eqs. 7 and 8) would be unlikely.
The hydrolyzing activity, however, seems to be too low to account
for the ADP-ribosylation activity. It might be plausible that the
binary complex between fragment A and NAD interacts normally
with EF 2 to form the transitory ternary complex, but occasionally
with H_2O to hydrolyze NAD.

Goor and Maxwell (*38*) made kinetic studies on various inhibitors
of the ADP-ribosylation reaction. They found that NMN and nicotinamide are competitive inhibitors with both NAD and EF 2,
whereas NADH and adenine derivatives are competitive only with
NAD, but not with EF 2. Based on these studies, they proposed
that toxin recognizes the adenine moiety of NAD and EF 2, while
the nicotinamide moiety is recognized by EF 2. Hayes and Kaplan
(*46*), however, showed strong binding of the nicotinamide portion
of NAD to fragment A by nuclear magnetic resonance studies. If
fragment A has NAD glycohydrolase activity, it would appear more
reasonable that fragment A should have a recognition site for nico-

tinamide. Further investigation seems to be required to elucidate the detailed mechanism of the reaction.

F. Linkage between ADP-Ribose and EF 2

The equilibrium constant for the ADP-ribosylation of EF 2 is calculated from Eq. (9), and the value of 6.3×10^{-4} was obtained from

$$K = \frac{\text{(ADP-ribosyl–EF 2) (nicotinamide) (H}^+)}{\text{(EF 2) (NAD)}} \qquad (9)$$

a series of experiments with varied concentrations of nicotinamide (*52*). It may readily be calculated from the equilibrium constant that the standard free energy change ($\Delta F^{\circ\prime}$) accompanying the ADP-ribosylation of EF 2 is about -5.2 kcal per mole at pH 7 and 25°C. The nicotinamide ribose linkage of NAD is a high-energy bond which contains about 9.2 kcal per mole at pH 7 and 25°C (*1, 121*). Therefore, the free energy of hydrolysis of the ADP-ribose EF 2 linkage is approximately 4.0 kcal/mole at pH 7 and 25°C as shown in Eqs. (10)–(12).

$$\text{NAD} + \text{EF 2} = \text{ADP-ribosyl–EF 2} + \text{nicotinamide} + \text{H}^+ - 5.2 \text{ kcal} \qquad (10)$$
$$\text{NAD} + \text{H}_2\text{O} = \text{nicotinamide} + \text{H}^+ + \text{ADP-ribose} - 9.2 \text{ kcal} \qquad (11)$$

Eq. (11) – Eq. (10.)

$$\text{ADP-ribosyl–EF 2} + \text{H}_2\text{O} = \text{ADP-ribose} + \text{EF 2} - 4.0 \text{ kcal} \qquad (12)$$

This value is close to the energies of hydrolysis reported for the ribosidic linkage of purine ribosides and the $\alpha(1 \to 4)$-glucosidic bond in the glycogen chain (about 4.0 and 4.3 kcal per mole, respectively (*2, 56*). It is reasonable to assume that in concentrations of NAD and EF 2 comparable to those found in living cells, the reaction goes almost to completion. However, the ADP-ribosylation reaction is accompanied by the release of a proton as shown in Eq (4), and therefore $\Delta F^{\circ\prime}$ accompanying the forward reaction is -2.5 kcal/mole at pH 5 and 25°C (*52*).

To date four kinds of amino acid residues are known to be linked to sugar moieties in various glycoproteins (*106*). They are (a) the hydroxyl group of serine or threonine, (b) the hydroxyl group of hydroxylysine or hydroxyproline, (c) the amidonitrogen of asparagine, and (d) the carboxyl group of glutamic acid or aspartic acid. The chemical stability of the ADP-ribosyl-EF 2 linkage was examined to see whether it would be similar to any of the above known linkages (53). The linkage is quite stable in 0.1 N NaOH at 50°C for 15 hours; under these conditions the O-glycosyl linkage attached to the hydroxyl group of serine or threonine is easily hydrolyzed (106). The rate constants of the hydrolysis of ADP-ribosyl EF 2 at 100°C in 1 N NaOH, 1 N HCl, and 0.2 N NaOH are 3.5 × 10^{-1} min^{-1}, 5.3 × 10^{-2} min^{-1}, and 4.8 × 10^{-2} min^{-1}, respectively. These results seem to exclude the possibility that the ADP-ribosyl linkage is on a hydroxyl group of hydroxylysine or hydroxyproline because the latter linkage is not hydrolyzed at 100°C in 1 N NaOH (106). The rate constant of hydrolysis of the N-glycosidic linkage to asparagine at 100°C in 1 N HCl is known to be 8 × 10^{-3} min^{-1} (66), which is one-seventh of the value for the ADP-ribosyl-EF 2 linkage. The ADP-ribosidic linkage resists hydroxylamine treatment (63), and so the fourth possibility, that it is attached to the carboxyl group of glutamic acid or aspartic acid, also seems unlikely.

In an attempt to investigate other possibilities, we have used various reagents to modify chemically the amino acid residues of EF 2 and to see whether these chemical modifications would affect the ability of EF 2 to accept the ADP-ribosyl moiety from NAD. We found that trinitrobenzene-sulfonate (TNBS), which is a specific modifier of the amino group (84), is very effective in diminishing the ADP-ribose accepting ability of EF 2. Treatment with acetic anhydride or succinic anhydride also inactivates this ability of EF 2. These results suggest that the amino group of EF 2 may be involved in the ADP-ribosyl linkage. In order to prove this hypothesis, we tried to isolate an ADP-ribosylated peptide or amino acid from ADP-ribosyl-EF 2 and to identify its chemical structure.

A large quantity of ADP-ribosylated EF 2 was prepared with NAD labeled with ^{14}C in the ribose portion of NMN as a starting material. After purification, it was subjected to treatment with trypsin and α-chymotrypsin. The radioactive fragment thus obtained was purified on a Dowex 50 column. The fragment was further treated with thermolysin and again purified on a Dowex 50 column. This fraction was subjected to high voltage paper electrophoresis at pH 5. As shown in Fig. 4, the label moved toward the cathode, indicating the basic nature of this component. When the

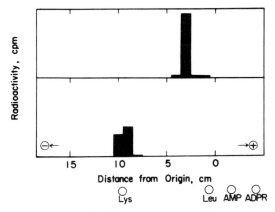

Fig. 4. Paper electrophoresis of ADP-ribosyl peptides and riboxyl peptides. Thermolysin-digested peptide was subjected to paper electrophoresis at 2000 V and pH 5.0 for 120 minutes (upper column). The peptide was further treated with phosphodiesterase and alkaline phosphatase, and then electrophoresed under the same conditions (lower column). Reproduced from Honjo *et al.* (*53*) by permission of Springer-Verlag, Berlin and New York.

component was further subjected to the action of phosphodiesterase and alkaline phosphatase, leaving only the ribose moiety on the peptide, the remaining fragment migrated to the cathode at a rate essentially similar to that of lysine (*53*). Further investigation is necessary to distinguish whether the ribosyl linkage is on a particular basic amino acid, or whether basic amino acids are simply in the vicinity of the ADP-ribosylated amino acid.

G. ADP-Ribosylation of EF 2 *in Vivo*

Diphtheria toxin has been shown to inhibit protein synthesis in advance of affecting RNA synthesis, aerobic glycolysis, oxygen uptake, or nucleotide metabolism in cultured cells (*61, 72, 107*). No effect of the toxin has been found to occur prior to the inhibition of protein synthesis, strongly suggesting that this represents the primary metabolic lesion of the toxin. Experiments in cell-free systems have established that ADP-ribosylation of EF 2 is responsible for the inhibition of protein synthesis by the toxin.

Studies on intoxicated whole animals or cultured cells have also shown that ADP-ribosylation of EF 2 takes place *in vivo* (*3, 5–8, 36, 71*). Baseman *et al.* (*3*) reported that various tissues from intoxicated guinea pigs or rabbits contain reduced amounts of EF 2 compared with those from normal animals. The decrease in the rate of protein synthesis is accompanied by a reduction in EF 2 content in various intoxicated organs such as the heart, kidney, pancreas, spleen, lung, and small intestine. In liver, however, the degree of inhibition of protein synthesis is much smaller than that of the reduction in EF 2 content. This might be due to the fact that liver contains an excess amount of EF 2 compared with other factors required for protein synthesis, as will be discussed later.

Supernatant fractions from intoxicated animals or cells contain inactivated EF 2 whose activity is restored by treatment with toxin and nicotinamide (*7, 8, 36*). The activity of polypeptide synthesis in 100,000 *g* supernatant fractions from heart or skeletal muscle of intoxicated guinea pigs is reactivated to the normal level only when assayed in the presence of toxin and nicotinamide (*7, 8*). Since the toxin is specific for EF 2, the results strongly suggest the presence of ADP-ribosylated EF 2 in the supernatant fractions from intoxicated tissues.

Bonventre and his co-workers (*6, 8*) reported that when low levels of toxin comparable to those which might be generated in a natural infection are given, inhibition of the soluble enzyme activity is found only in extracts from heart and skeletal muscle. At high toxin dosages, the inhibition is widespread in various other organs, such as lung, liver, spleen, and kidney in addition to muscle tissues. Baseman *et al.* (*3*), however, reported that by the time a test animal is moribund some reduction in EF 2 and in protein synthesis is apparent in every organ tested although not all organs respond to the toxin equally rapidly or to an equal extent. Protein synthesis in tissue slices from intoxicated animals was measured at various time intervals after an intravenous injection of low doses of the toxin. In several tissues (group I), leucine uptake began to decrease within 2–3 hours and was reduced linearly with time. Group I tissues include kidney and various types of muscle tissues. In other tissues (group II), including liver, lung, pancreas, and spleen, there was a pronounced lag of several hours up to 16 hours before a significant decrease in leucine uptake was observed. This might mean that group II tissues contain EF 2 in relative excess and that group I tissues have a limited amount of EF 2 compared with other factors required for protein synthesis. Alternatively, group II tissues may be resistant to the attack of the toxin for some other reasons. The presence of a pronounced lag in group II tissues may provide a possible explanation for the above-described observation by Bonventre *et al.* (*6, 8*). These investigators also detected inactivated EF 2 only in extracts from heart and other muscle tissues with the use of the reverse reaction (*7, 8*). From these observations they proposed that ADP-ribosylation of EF 2 occurs primarily if not exclusively in muscle tissues of intoxicated animals and that the impairment of protein synthesis in other tissues might reflect a secondary effect engendered during the progression of the toxemia. One has to take into account, however, that the reverse reaction is active only in the pH range between 5 and 7 (*50*). In addition, preincubation is preferably to ensure the completion of the reverse reaction. Bonventre and his co-workers (*7, 8*) did not disclose the precise pH of the assay system, but they probably worked at approximately pH 7.5, since this is the pH for the assay of protein synthesis. They did not preincubate the supernatant fractions with the toxin and nicotinamide. Further investigation will be necessary to evaluate their proposal.

In conclusion, even though secondary lesions undoubtedly contribute to the symptoms of diphtheria, it is clear that ADP-ribosylation of EF 2 is the primary effect of the toxin, and that the resulting inhibition of protein synthesis leads ultimately to the death of the animals.

H. Structure of Toxin and ADP-Ribosylating Activity

Diphtheria toxin is a protein with a molecular weight of 60,000 to 70,000 and a sedimentation constant of 4.2–4.6 S (*60, 92*). The amino acid composition and terminal residues of the crystalline preparation of the toxin have been reported. The toxin contains 2 disulfide bonds but no free SH groups. Kato (*60*) found one glycine and 0.45 serine as NH_2-terminal of the toxin molecule and one glycine, one serine and one alanine as COOH-terminal. Diphtheria toxin has therefore been considered to be a protein consisting of at least two peptide chains linked together by disulfide bonds. Treatment of the toxin molecule with sodium bisulfate and copper in 8 M urea splits these disulfide bonds and yields two different peptides, neither of which is toxic to sensitive animals (77).

The discovery of the ADP-ribosylation activity of the toxin prompted reinvestigation of the structure and activity relationship of this molecule. Collier and his co-workers (16, 19) have shown that DTT treatment dissociates the toxin into 2.5 S subunits which have no toxicity to sensitive animals but can ADP-ribosylate EF 2. The total catalytic activity increases more than 7-fold by DTT treatment. The catalytic activity of the toxin is also activated about 10-fold upon heat treatment at alkaline pH, under which conditions the toxicity is completely destroyed (52). This activation is accompanied by dissociation of the toxin molecule (4.2 S) into subunits with sedimentation coefficients of 2.5 S.

Gill and Dinius (32) have analyzed various preparations of diphtheria toxin on polyacrylamide gel electrophoresis in the presence of sodium dodecyl sulfate (SDS). Nearly all preparations consisted predominantly of a protein with a molecular weight of about 62,000. However, when disulfide bonds were reduced by 2-mercaptoethanol or dithiothreitol (DTT), SDS gel analysis revealed significant differences between the preparations. Some preparations were apparently unaltered by reduction, while in one case the 62,000-dalton material broke down almost entirely to equiomolar amounts of two peptides of molecular weights of 24,000 (fragment A) and 38,000 (fragment B). Most preparations were mixtures of the 62,000-dalton component, and fragments A and B. Sometimes, three or more kinds of peptides were obtained after reduction. Clearly some toxin preparations possess "intact toxin" consisting of a single peptide chain and others contain "nicked toxin" consisting of two disulfide-linked polypeptide chains. Fragment A contains one half-cystine, and fragment B contains three (23, 32). Since the intact or nicked toxin contains two disulfide bonds, fragment A is linked to fragment B through a single disulfide bond and the other disulfide bond is on fragment B. Collier and Kandel (17) also obtained essentially the same results except that they estimated the molecular weight of fragment B to be 39,000.

The intact toxin exhibits very little or no ADP-ribosylation activity toward EF 2 (*17*, *23*, *34*). The nicked toxin also shows little catalytic activity unless it is reduced by sulfhydryl reagents. However, it is difficult to conclude that the intact or nicked toxin has no catalytic activity at all, because no toxin preparation is completely free of contamination by fragments A and B. Fragment A has the ADP-ribosylating activity, the specific activity of isolated fragment A being 2.4 times as great as that of the reduced nicked toxin. The molecular weight of the nicked toxin is 2.6 times as large as fragment A and the close coincidence makes it likely that the entire enzymatic activity of the nicked toxin molecule can be attributed to fragment A. The presence of fragment B does not affect the reaction rate. Although fragment B is highly unstable and no one has measured the catalytic activity of purified fragment B, available evidence indicates that fragment B is catalytically inactive. Further evidence to support this idea was provided by the experiments using mutant toxin proteins (*115*, *116*). The structure of the toxin is summarized in Fig. 5.

Gentle treatment with several proteases such as trypsin, Pronase, and subtilisin converts the intact toxin into partially digested toxin which is indistinguishable from naturally obtained nicked toxin by the following criteria (*23*, *34*). (a) The digested toxin shows the same mobility as that of the nicked toxin on SDS polyacrylamide gel electrophoresis. (b) The digested toxin is catalytically as inactive as the nicked toxin unless reduced by sulfhydryl reagents. (c) Upon reduction with sulfhydryl reagents, the digested toxin

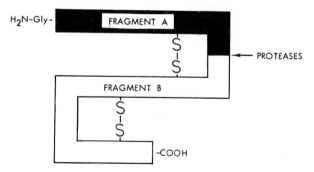

FIG. 5. Structure of diphtheria toxin.

yields two peptides corresponding to fragments A and B on SDS gel elctrophoresis. (d) Fragment A derived from digested toxin has the same specific activity for ADP-ribosylation as that of the fragment from the nicked toxin. Although it is premature to conclude that digested toxin is identical to nicked toxin, they seem to be closely related to each other. It is very likely that the bacteria produce the intact toxin, which is then cleaved at a specific site on the polypeptide chain by proteases present in the bacterial culture medium. Indeed, such protease activity has been detected in a crude preparation of the toxin (*32*). The ADP-ribosylating activity of the toxin, which had been demonstrated before the detailed structure of the toxin was known, seems to be largely attributable to fragment A. Since the sulfhydryl reagent was an obligatory ingredient of the buffer solution to preserve EF 2, the nicked toxin, which is found in almost all the toxin preparations, must have been unknowingly reduced to give rise to fragment A when the toxin was mixed with NAD and EF 2.

As discussed earlier, it is certain that ADP-ribosylation of EF 2 is responsible for the toxicity of diphtheria toxin. Why then does catalytically active fragment A have no toxicity at all? Gill and Pappenheimer (*34*) and Drazin et al. (*23*) proposed that fragment A per se is unable to get into the cytoplasm, whereas the intact or nicked toxin can. Fragment B or that portion of the intact toxin corresponding to fragment B may act to bind the toxin molecule to a specific site on the cell membrane and/or to facilitate transport of the fragment A into the cell. Recent reports by Uchida et al. (*115, 116*) provided evidence to support this hypothesis. They isolated several mutants of β phase which direct synthesis of proteins immunologically related to the toxin in lysogenic bacterial cells. These reports, incidentally, provided unequivocal evidence that the structural gene of the toxin is coded by the β phage genome. In Table IV, some properties of four such cross-reacting proteins are compared with the properties of the wild-type toxin and its fragments. Crm_{45} and crm_{30} consist of single polypeptides with molecular weights of 45,000 and 30,000, respectively. They are catalytically active although the specific activity is considerably lower than that of fragment A. A polypeptide indistinguishable from fragment A on SDS gel electrophoresis and also in its catalytic activity is gen-

TABLE IV

SOME PROPERTIES OF DIPHTHERIA TOXIN AND RELATED PROTEINS[a]

| Protein | Molecular weight | ADP-ribosylating activity[b] | | Toxicity[c] |
		Intact	Activated	
Toxin	62,000	0	100	100
Fragment A	24,000	—	100	0
Fragment B	38,000	—	0	0
crm_{45}	45,000	Partially active	100	0
crm_{30}	30,000	Partially active	100	0
crm_{197}	62,000	0	0	0
crm_{176}	62,000	0	10	<0.4

[a] Data were taken from Uchida et al. (116) and Gill et al. (37).
[b] Expressed as percent of activity of fully activated (nicked and reduced) toxin on a molar basis.
[c] Expressed as percent of intradermal toxicity of purified toxin.

erated when crm_{45} or crm_{30} is digested gently with trypsin and then reduced. These results would indicate that crm_{45} and crm_{30} carry a chain terminating or deleting mutation somewhere in the toxin gene coding for fragment B, but not in the region coding for fragment A. It follows, therefore, that fragment A may be located on the amino end of the toxin molecule. The presence of glycine at NH_2 termini of both toxin and fragment A (69) is consistent with this supposition. Crm_{197} and crm_{176} have the same molecular weight as the toxin. Crm_{197} has no enzymatic activity, even if treated with protease and an SH reducing agent. Crm_{197} does not show any toxicity at all. On the other hand, crm_{176} is 10% as active as the native toxin when partially digested and reduced. Crm_{176} is 0.4% as toxic as the intact toxin. It may be concluded that crm_{197} and crm_{176} have missense mutations somewhere in the fragment A region. Uchida et al. (116) have proved elegantly that crm_{197} has a normal fragment B. They have reconstituted the nicked toxin from crm_{197} and crm_{45}, indicating that crm_{197} and crm_{45} contain functionally complementary portions of the native toxin. Since crm_{45} lacks a portion of fragment B, there is no doubt that crm_{197} has functionally active fragment B in its structure. Uchida et al. (116) have also shown that crm_{197} has a very interesting property which suggests that fragment B may have a specific role in binding the toxin molecule to specific sites on the cell membrane. As illustrated in Fig. 6, increasing the amount of crm_{197} competitively blocked the toxic activity of the native toxin toward HeLa cells. On the other hand, crm_{45} or toxoid was unable to prevent inhibition of protein synthesis by the toxin. The apparent association constant of crm_{197} for a surface receptor is of the order of 10^8 liters per mole (37). From these results it seems likely that fragment B is essential for the toxin to attach to the HeLa cell membrane. However, further investigation is necessary to determine whether fragment B per se has the attachment site to the receptor on the cell membrane or the fragment is required to maintain a specific conformation of the toxin molecule whereby it can bind to the receptor on the cell membrane. There are no experimental results to indicate that fragment B facilitates the entry of fragment A into the cytoplasm. From radioautographic studies on HeLa cells treated with ^{125}I-labeled diphtheria toxin, Pappenheimer and Brown (87) reported that a small amount of the toxin becomes associated with HeLa cells and that most of this labeled protein is found near the periphery of the cells. Gill and Pappenheimer (34) argued without any evidence that this peripheral protein might have been largely fragment B left behind on the cell membrane, and that the small amount of protein present within the HeLa cell might have been fragment A. It is still unknown, however, whether both fragments A and B enter the cytoplasm or not. It remains very important to elucidate the function of fragment B, since this may provide the key leading to an answer to the question, "How do insensitive cells differ from sensitive cells?"

I. Applications of ADP-Ribosylation of EF 2

Since the ADP-ribosylation reaction is specific for EF 2, the reaction has been utilized to clarify the function of EF 2 in polypeptide synthesis. Schneider et al. (96) showed that in a cell-free system from rabbit reticulocytes translocation of peptidyl-tRNA on ribosomes is inhibited by the presence of the toxin and NAD. Raeburn et al. (91) and Honjo et al. (50) reported that in the presence of NAD the toxin inactivates the ribosome-dependent GTPase activity in mammalian systems. Thus it is clear that both ribosome-depen-

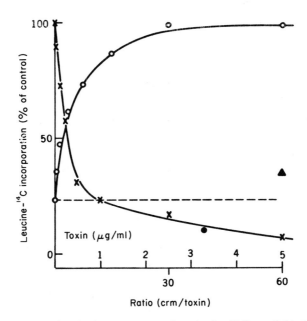

FIG. 6. Competition between crm₁₉₇ and toxin for HeLa cell binding sites. Washed suspensions of growing HeLa cells were suspended in Eagle's medium containing 2% fetal calf serum and then distributed in 2-ml amounts in roller tubes containing increasing amounts of purified crm₁₉₇. After 30 minutes' rotation at 5 rpm at 37°C, diphtheria toxin (1 μg/ml) was added to all tubes, except for certain control tubes. After incubation for 3.5 hours more, 100 μl of leucine-¹⁴C (1 μCi/ml) was added to each tube. Cells were harvested 3.5 hours later on Millipore filters, washed first with Hanks salt solution, then with 5% trichloroacetic acid, dried, and counted. Leucine uptake by control tubes without toxin was taken as 100%. In the presence of 60 μg of crm₁₉₇ alone, leucine incorporation was the same as in the control. The open circles show the effect of increasing ratios of crm₁₉₇ to toxin on leucine-¹⁴C uptake plotted as percentage of the control uptake. The crosses represent titration of toxin under the same experimental conditions. Incorporation of leucine in the presence of toxin (1 μg/ml) alone was 23% of the control value (dashed line). The filled circle and triangle show leucine uptake in the presence of crm₄₅ and purified toxoid in ratios to toxin of 40:1 and 60:1, respectively. Reproduced from Uchida et al. (116) by permission of the American Association of the Advancement of Science.

pendent GTPase and translocase activities in mammalian systems can be ascribed to EF 2. It follows therefore that EF 2 corresponds functionally to G factor or EF G in bacterial systems, although the latter enzyme is not affected at all by the toxin.

Since 1 mole of ADP-ribose binds to 1 mole of EF 2, the ADP-ribosylation reaction can provide a measure of the amount of EF 2 in the presence of excess amount of isotopically labeled NAD (20, 33, 52, 89). By this method Gill et al. (35) quantitated the number of EF 2 molecules which bind to a ribosome using the fact that any EF 2 bound to ribosomes is inert as a substrate of the ADP-ribosylation reaction. They concluded that 6 moles of EF 2 bind to 1 mole of ribosomes. Bound EF 2 is released from ribosomes by various guanine nucleotides such as GTP, GDP, and GMPPCP. Traugh and Collier (113, 114) found that EF 2 forms a complex with ribosomal RNA and the interaction is inhibited by GTP and GDP. These findings appear contradictory to the report that GTP is required for binding of EF 2 to ribosomes when assayed by the activity of polypeptide synthesis (100). One possible explanation for these observations is as follows: A ribosome has one functional site and six resting sites for EF 2 binding. GTP is necessary for binding to the functional site whereas the nucleotide releases EF 2 from the resting sites. Several reports seem to indicate that most of EF 2 molecules in vivo bind to resting sites on monosomes and presumably to functional sites on polysomes (102, 103). Smulson et al. (102, 103) employed essentially the same technique and showed that in HeLa cells the ratio of the amount of free EF 2 to that of EF 2 bound to ribosomes varies under different nutritional conditions.

III. ADP-Ribosylation of Nuclear Proteins

A. Poly-ADP-Ribose Synthesis

Mammalian nuclei have been shown to catalyze polymerization of the ADP-ribose moiety of NAD into a novel homopolymer with repeating ADP-ribose units (13, 25, 80, 93, 109). The structure of the polymer, which is referred to as poly-ADP-ribose, has been established as shown in Fig. 7 (13, 22, 45, 82, 93, 98). ADP-ribose units are linked together through a ribosyl–ribose linkage. Snake

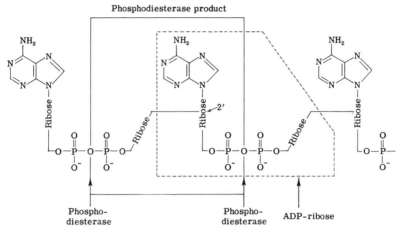

FIG. 7. Structure of poly-ADP-ribose. Reproduced from Nishizuka *et al.* (*82*) by permission of the Cold Spring Harbor Laboratory.

venom phosphodiesterase splits the pyrophosphate bond to yield a product identified as 2'-(5''-phosphoribosyl)-5'-AMP.

Poly-ADP-ribose shows the general characteristics of a polynucleotide. It precipitates in 5% trichloroacetic acid or in 66% ethanol containing 0.2 M acetate buffer, pH 4.8. On sucrose density gradient centrifugation poly-ADP-ribose has a broad size distribution with a peak at 4 S (*13, 45, 93*). It has a buoyant density of 1.57 on cesium sulfate equilibrium density gradient centrifugation (*45*). On methylated albumin kieselguhr the polymer also shows a broad distribution (*93*). On a hydroxyapatite column, poly-ADP-ribose is eluted after RNA and DNA at a much higher concentration of buffer (*111*). Extensive purification of this polymer has been accomplished by gel filtration (*97*) or by hydroxyapatite column chromatography (*111*).

The enzyme activity to synthesize poly-ADP-ribose is distributed in many tissues of various vertebrates (*47, 81, 88*). The yeast *Saccharomyces cerevisiae* (*110*) and the protozoan *Tetrahymena pyriformis* (*108*) also have this activity. However, no activity has been found either in higher plants or in prokaryotic organisms (*81*). The enzyme activity is strictly localized in the nucleus with approximately 90% of the total activity associated with the chromatin. In addition, the activity of the ADP-ribosylation reaction appears tightly associated with DNA (78, 82, 118) since it is significantly inhibited when the chromatin is digested with DNase but not with RNase (13, 80, 82). The average chain length of the polymer varies from 6 to 30 depending on the nuclear preparation (82, 98). Gentle treatment of the chromatin with DNase reduces the average chain length (82). The apparent optimal pH for poly-ADP-ribose synthesis is 8.0 with Tris buffer (25, 80) and 8.6 with glycine buffer (13). The reaction is specific for NAD, and NADP and deamido-NAD are inert as substrates. The acetylpyridine analog of NAD is only 4% as active as NAD (76). Nicotinamide strongly inhibits the reaction but is unable to reverse it (25). p-Chloromercuribenzonate (13, 25, 80), 5-methylnicotinamide (14), NMN (88), thymidine (88), bromodeoxyuridine (88), or iododeoxyuridine (88) also inhibit the formation of poly-ADP-ribose. Many antibiotics such as actinomycin D, mitomycin C, chloramphenicol, cycloheximide, and chromomycin A_3 do not inhibit the reaction (25).

Ueda et al. (118) have shown that the enzymatic activity is greatly influenced by the presence of various salts. It is interesting to note that low concentrations (0.1–0.5 M) of ammonium sulfate markedly depress poly-ADP-ribose synthesis in contrast with a remarkable increase in RNA synthesis. However, at higher concentrations, poly-ADP-ribose synthesis increases again and reaches a peak at 1.7 M when synthesis is 40% as active as in the salt-free state.

The poly-ADP-ribose synthesized remains in association with deoxyribonucleoprotein when subjected to gel filtration in the presence of salt (118), but is dissociated from DNA in a cesium sulfate density gradient. SDS or proteases prompted the release of poly-ADP-ribose from DNA, suggesting that the polymer is linked to DNA through a protein molecule.

Radioautographic analysis has shown that poly-ADP-ribose is evenly distributed over the extranucleolar region of nuclei of Ehrlich ascites tumor cells (47) and of rat embryonic cells (83).

B. ADP-Ribosylation of Nuclear Proteins

Nishizuka et al. (79, 82) and Otake et al. (86) subsequently showed that poly-ADP-ribose is attached to nuclear proteins, prob-

TABLE V
FRACTION OF NUCLEAR PROTEINS AND POLY-ADP-RIBOSE[a,b]

Fraction	Radioactivity insoluble in 20% trichloracetic acid		Specific activity (cpm/mg protein)
	Cpm	%	
Whole nuclei	63,700	100	2,353
Globulins	9,500	15	799
Histones	46,560	73	4,865
Residual proteins	4,800	8	774

[a] Reproduced from Nishizuka et al. (79) by permission of the American Society of Biological Chemists, Inc.

[b] The reaction mixture (2.5 ml), containing 36 nmoles of NAD-(adenine-8)[14]C (476,000 cpm), freshly prepared nuclei (27.5 mg of protein), 250 nmoles of Tris·HCl buffer, pH 8.0, and 75 μmoles of $MgCl_2$, was incubated for 10 minutes at 37°C. The mixture was chilled to 0° and immediately extracted twice, each time with 20 ml of 0.01 M Tris·HCl, pH 7.4 containing 3.3 mM $CaCl_2$. Histones were subsequently extracted five times, each time with 10 ml of 0.25 N HCl by homogenizing the pellet with a glass and Teflon homogenizer followed by centrifugation for 20 minutes at 25,000 g.

ably histones. After incubation of rat liver nuclei with NAD-(adenine)-^{14}C, the radioactive nuclear preparation was fractionated into various nuclear proteins as shown in Table V. Acid-insoluble radioactivity was found to be distributed throughout all nuclear protein fractions. However, approximately 75% of the radioactivity was recovered in the dilute HCl-extractable fraction which contains predominantly histones (79, 82). The radioactive material in the latter fraction was associated with all histone subfractions upon chromatography on a carboxymethyl-cellulose column (79, 82, 86). This result, however, does not exclude the possibility that the ADP-ribose units are attached to some other proteins which contaminate the histone fraction.

The radioactive product associated with proteins shows a size distribution from monomers up to polymers with molecular weights of at least several thousands (82). These results indicate that the ADP-ribose portion of NAD is transferred to histones and a successive transfer of ADP-ribose units results in the formation and elongation of the polymer. It is plausible that histones serve as a primer of poly-ADP-ribose synthesis and are simultaneously modified by the enzymatic reaction.

Poly-ADP-ribose per se, when free of protein, is alkali stable. The ribosyl linkage between the ribose of the first ADP-ribose unit and protein, however, is extremely labile under alkaline conditions, although it resists acid treatment (82). That the half-life of the ADP-ribosyl histone linkage in 0.1 N NaOH at 0°C is approximately 5 minutes makes it unlike any of the previously known glycopeptides such as N- or O-glycosides attached to asparagine, serine, or hydroxylysine residues of protein molecules. The ADP-ribosyl linkage to nuclear proteins reacts with neutral hydroxylamine (82), a reagent that has been well established to react with acyl anhydride (63). The ADP-ribose units attached to nuclear proteins are removed by hydroxylamine treatment as rapidly as is phenylalanine esterified to tRNA. The ADP-ribosyl linkage to EF 2 does not react at all. These results support the proposal that the carboxyl group in the protein molecule is the residue to which the ADP-ribose moiety is transferred (82).

C. Enzymes of Poly-ADP-Ribose Metabolism

1. ADP-RIBOSE TRANSFERASE

ADP-ribose transferase, which catalyzes the formation of poly-ADP-ribose, has been also referred to as ADP-ribose polymerase (26) or poly-ADP-ribose polymerase (43). Its enzymatic activity is highly sensitive to DNase treatment and to the ionic concentration of the reaction mixture. The effect of these two conditions upon poly-ADP-ribose synthesis is very different (116a). Incubation with DNase results in substantial decrease in the average chain length of poly-ADP-ribose synthesized. In contrast, varying the ionic strength of the reaction mixture increases or decreases the number of polymer chains. These observations suggest that it may be possible to distinguish the activity required for the polymer chain initiation from that required for its elongation (116a).

The tight association with nucleoprotein has prevented the purification and characterization of this enzyme. With DNase treatment, the enzyme has been partially dissociated from DNA (99) and further purified 10-fold from rat liver nuclei by ammonium sulfate fractionation (119). Recently, Yoshihara (120) reported that the enzyme was purified 130-fold from rat liver and obtained essentially free of DNA. The enzyme was separated from DNA by CsCl density gradient centrifugation in the presence of 30% glycerol and was further purified by hydroxyapatite and carboxymethyl cellulose column chromatography. The enzyme activity is absolutely dependent on DNA (119, 120). Although a single-stranded DNA (phage ϕX174), as well as double-stranded DNA's, is effective, rat liver DNA is the most effective. No specific base sequences in DNA, however, seem to be prerequisite, since all synthetic deoxyribonucleotide polymers and DNA–RNA hybrids so far tested, except poly[d(T)], can replace natural DNA in this reaction (120). Actinomycin D, ethidium bromide, and proflavin inhibit the reaction to the extent of 90%, 40%, and 40%, respectively, at a concentration of 8 μg/ml (120). The purified enzyme shows a partial requirement for histones (119, 120). The reaction rate is stimulated 2- to 3-fold by the addition of calf thymus histones whereas bovine serum albumin gives no stimulation. The absence of an absolute requirement for exogenous histones may be due to histones or other nuclear proteins contaminating the enzyme preparation. The enzyme activity is also dependent on DTT and divalent cations such as Mg^{2+}, Ca^{2+}, and Mn^{2+} (119). The molecular weight of the enzyme is estimated to be 78,000 by sucrose density gradient centrifugation (119). Gill (31) obtained an apparently soluble preparation of ADP-ribose transferase by direct extraction of tissue homogenates with 0.5 M NaCl. This enzyme preparation, after storage for several days at $-20°$, shows some stimulation by the addition of double-stranded DNA.

2. PHOSPHODIESTERASE

Phosphodiesterase from rat liver has been shown to cleave the pyrophosphate bonds of poly-ADP-ribose to yield 2′-(5″-phosphoribosyl)-5′-AMP and 5′-AMP from the terminus of the chain (27–29). This enzyme, which also hydrolyzes p-nitrophenyl uridine 5′-monophosphate, various oligonucleotides and NAD, has been purified from the particulate fraction of rat liver (28) and is bound to a nucleonemata structure (117). Its optimal pH is approximately 10 (28). This phosphodiesterase hydrolyzes poly-ADP-ribose from the AMP terminus in an exonucleolytic fashion (68). In contrast, snake venom phosphodiesterase digests poly-ADP-ribose in an endonucleolytic fashion (67). Poly-ADP-ribose is not hydrolyzed by pancreatic DNase, pancreatic RNase, RNase T_1, spleen phosphodiesterase, potato phosphodiesterase micrococcal nuclease, or potato nucleotide pyrophosphatase (13, 25, 45, 80, 93, 109).

3. POLY-ADP-RIBOSE GLYCOHYDROLASE

Poly-ADP-ribose glycohydrase which hydrolyzes the ribose–ribose linkage of poly-ADP-ribose has been recently demonstrated in crude nuclear preparations from calf thymus (70) and rat liver (117). Ueda et al. (117) purified this enzyme so that it was essentially free of ADP-ribose transferase and phosphodiesterase. The enzyme catalyzes hydrolytic cleavage of the ribosyl $(1 \to 2)$ ribose bond between neighboring ADP-ribose units, but does not split the linkage whereby the polymer is bound to protein. Neither does it hydrolyze 2'-(5''-phosphoribosyl)-5'-AMP, a phosphodiesterase-digested product of poly-ADP-ribose. The enzyme is found in two states; one, soluble in nucleoplasm and the other, associated with chromatin (70a). Its association with chromatin, however, is not very tenacious, since about half of the bound enzyme is solubilized with 0.15 M NaCl. Glycohydrolase is most active at pH 6.5–7 and is sensitive to high ionic strength; $(NH_4)_2SO_4$ and NaCl inhibit the activity by 47 and 38%, respectively, at 0.1 M (70a). Adenosine 3',5'-cyclic monophosphate inhibits the glycohydrolase activity with an apparent K_i value of 1.5 mM. Kinetic studies and product analyses in crude preparations have suggested that glycohydrolase, rather than phosphodiesterase, plays a principal role in the degradation of poly-ADP-ribose (31, 70, 117).

4. NADASE

Mammalian nuclei have an NADase activity distinct from microsomal NADase (4, 76, 81, 94, 112). Comparative studies have revealed that nuclear NADase and ADP-ribose transferase show very similar properties. The possibility that ADP-ribose transferase might be one functional aspect of the transglycosidase activity of NADase present in the nucleus has been extensively discussed (4, 76, 81, 94, 110).

D. Possible Biological Function

1. Occurrence in Nature

Poly-ADP-ribose is demonstrated in chicken liver labeled with $^{32}P_i$ *in vivo* (*21*). In mammalian systems no evidence has yet been presented to indicate the natural occurrence of poly-ADP-ribose synthesis. ADP-ribosylation of histones *in vivo* has not been reported in any organism.

2. Regulation of DNA Synthesis

Burzio and Koide (*10*) reported that preincubation of rat liver nuclei or chromatin with NAD results in a depression of the incorporation of ^3H-TTP into DNA. When nuclei are preincubated with increasing concentration of NAD, the amount of poly-ADP-ribose synthesized increases, reaching a plateau of 2 mM NAD. DNA synthesis in preincubated nuclei is reduced in inverse proportion to poly-ADP-ribose synthesis, while RNA synthesis is unaffected (*11*). The addition of nicotinamide in the preincubation mixture prevents the inhibition of DNA synthesis. This observation has been confirmed in regenerating liver (*12*) slime mold *Physarum polycephalum* (*9*), Ehrlich carcinoma cells (*48*), and HeLa S3 cells (*48*). Treatment of nuclei from Novikoff hepatoma with NAD, however, does not alter the template activity for DNA synthesis (*12*). Preincubated chromatin shows decreased template activity when assayed by exogenous DNA polymerase from rat liver or *Micrococcus lysodeikticus* (*10, 11*). The addition of calf thymus DNA to the preincubated chromatin partially restores the DNA-synthesizing activity. DNA obtained from chromatin preincubated with NAD shows template activity equivalent to that from untreated chromatin (*11*). It was proposed from these observations that ADP-ribosylation of histones results in a depression of the template capacity of chromatin for DNA synthesis. Nagao et al. (*75*), however, reported that DNA polymerase per se is inactivated when rat liver nuclei are preincubated with NAD. In contrast to the findings of Burzio and Koide (*10, 11*), DNA synthesis in the preincubated chromatin is not restored by the addition of poly-d(A-T). In addition, the template activity of inactivated chromatin is identical to the untreated control when assayed with *E. coli* DNA polymerase.

Smulson *et al.* (*101*) found that ADP-ribose transferase has maximal specific activity during G_1 and a significantly lower specific activity during the S phase of the HeLa cell cycle. This suggests that ADP-ribosylation of nuclear proteins might be an important nuclear event at some time immediately before DNA replication.

Hilz and Kittler (*48*) cast doubt on the theory that ADP-ribosylation of nuclear proteins plays a regulatory role in DNA synthesis (*10–12*). They found no correlation between the poly-ADP-ribose transferase activity and the rate of DNA synthesis in normal, neonatal, and malignant cells. Instead, the enzyme activity is in parallel with the DNA content in normal, neonatal, and malignant cells. They suggested that poly-ADP-ribose might have a structural role within the nucleus.

Haines *et al.* (*43*) measured the ADP-ribose transferase activity in various fractions of rat liver nuclei separated by zonal centrifugation (*54, 55*). The highest activity of the enzyme is found in nuclei involved in RNA synthesis rather than DNA synthesis. They discussed the possibility that poly-ADP-ribose is a reservoir of NAD to maintain a constant level of intracellular NAD. But the reverse reaction to form NAD has not yet been described (*80, 108*). Experimental results using 5′-methyl nicotinamide, an inhibitor of ADP-ribose transferase, do not favor this theory (*14*).

Extensive efforts are being maintained to elucidate the biological function of ADP-ribosylation of histones. However, in view of the contradictory experimental data, it is clear that considerable work is yet required before the biological function of this reaction will be understood.

ACKNOWLEDGMENTS

The authors are grateful to Drs. D. D. Brown, P. M. Lizardi, R. H. Reeder, P. J. Stambrook, K. Ueda, Mr. R. Stern, and Miss M. Ohara for their critical reading of this manuscript and to Drs. M. Gill, E. Maxwell, and T. Sugimura for sending manuscripts prior to publication.

REFERENCES

1. Atkinson, M. R., Johnson, E., and Morton, R. K., *Biochem. J.* **79,** 19 (1961).

2. Barton, K., and Krebs, H. A., *Biochem. J.* **54**, 94 (1953).
3. Baseman, J. B., Pappenheimer, A. M., Jr., Gill, D. M., and Harper, A. A., *J. Exp. Med.* **132**, 1138 (1970).
4. Bock, K. W., Gang, V., Beer, H. P., Kronaw, R., and Grunicke, H. *Eur. J. Biochem.* **4**, 357 (1968).
5. Bonventre, P. F., and Imhoff, J. G., *J. Exp. Med.* **124**, 1107 (1966).
6. Bonventre, P. F., and Imhoff, J. G., *J. Exp. Med.* **126**, 1079 (1967).
7. Bowman, C. G., and Bonventre, P. F., *Biochem. Biophys. Res. Commun.* **41**, 1148 (1970).
8. Bowman, C. G., and Bonventre, P. F., *J. Exp. Med.* **131**, 659 (1970).
9. Brightwell, M., and Shall, S., *Biochem. J.* **125**, 67p. (1971).
10. Burzio, L., and Koide, S. S., *Biochem. Biophys. Res. Comun.* **40**, 1013 (1970).
11. Burzio, L., and Koide, S. S., *Biochem. Biophys. Res. Commun.* **42**, 1185 (1971).
12. Burzio, L., and Koide, S. S., *FEBS Lett.* **20**, 29 (1972).
13. Chambon, P., Weil, J. D., Doly, J., Storosser, M. T., and Mandel, P., *Biochem. Biophys. Res. Commun.* **25**, 634 (1966).
14. Clark, J. B., Ferris, G. M., and Pinder, S., *Biochim. Biophys. Acta* **238**, 82 (1971).
15. Collier, R. J., *J. Mol. Biol.* **25**, 83 (1967).
16. Collier, R. J., and Cole, H. A., *Science* **164**, 1179 (1969).
17. Collier, R. J., and Kandel, J., *J. Biol. Chem.* **246**, 1497 (1971).
18. Collier, R. J., and Pappenheimer, A. M., Jr., *J. Exp. Med.* **120**, 1019 (1964).
19. Collier, R. J., and Traugh, J. A., *Cold Spring Harbor Symp. Quant. Biol.* **34**, 589 (1969).
20. Collins, J. F., Raeburn, S., and Maxwell, E. S., *J. Biol. Chem.* **246**, 1049 (1971).
21. Doly, J., and Mandel, P., *C. R. Acad. Sci.* **264**, 2687 (1967).
22. Doly, J., and Petek, F., *C. R. Acad. Sci.* **263**, 1341 (1966).
23. Drazin, R., Kandel, J., and Collier, R. J., *J. Biol. Chem.* **246**, 1504 (1971).
24. Everse, J., Gardner, D. A., Kaplan, N. O., Galasinski, W., and Moldave, K., *J. Biol. Chem.* **245**, 899 (1970).
25. Fujimura, S., Hasegawa, S., Shimizu, Y., and Sugimura, T., *Biochim. Biophys. Acta* **145**, 247 (1967).
26. Fujimura, S., and Sugimura, T., *in* "Methods in Enzymology" (D. B. McCormick and L. D. Wright, eds.), Vol. 18B, p. 223. Academic Press, New York, 1971.
27. Futai, M., and Mizuno, D., *J. Biol. Chem.* **242**, 5301 (1967).
28. Futai, M., Mizuno, D., and Sugimura, T., *Biochem. Biophys. Res. Commun.* **28**, 395 (1967).
29. Futai, M., Mizuno, D., and Sugimura, T., *J. Biol. Chem.* **243**, 6325 (1968).

124 TASUKU HONJO AND OSAMU HAYAISHI

30. Galasinski, W., and Moldave, K., *J. Biol. Chem.* **244**, 6527 (1969).
31. Gill, D. M., *J. Biol. Chem.* **247**, 5964 (1972).
32. Gill, D. M., and Dinius, L. L., *J. Biol. Chem.* **246**, 1485 (1971).
33. Gill, D. M., and Dinius, L. L., *J. Biol. Chem.* **248**, 654 (1973).
34. Gill, D. M., and Pappenheimer, A. M., Jr., *J. Biol. Chem.* **246**, 1492 (1971).
35. Gill, D. M., Pappenheimer, A. M., Jr., and Baseman, J. B., *Cold Spring Harbor Symp. Quant. Biol.* **34**, 595 (1969).
36. Gill, D. M., Pappenheimer, A. M., Jr., Brown, R., and Kurnick, J. T., *J. Exp. Med.* **129**, 1 (1969).
37. Gill D. M., Pappenheimer, A. M., Jr., and Uchida, T., *Fed. Proc.*, *Fed. Amer. Soc. Exp. Biol.* **32**, 1508 (1973).
38. Goor, R. S., and Maxwell, E. S., *J. Biol. Chem.* **245**, 616 (1970).
39. Goor, R. S., and Pappenheimer, A. M., Jr., *J. Exp. Med.* **126**, 899 (1967).
40. Goor, R. S., and Pappenheimer, A. M., Jr., *J. Exp. Med.* **126**, 913 (1967).
41. Goor, R. S., Pappenheimer, A. M., Jr., and Ames, E., *J. Exp. Med.* **126**, 923 (1967).
42. Gumport, R. I., and Lehman, I. R., *Proc. Nat. Acad. Sci. U.S.* **68**, 2559 (1971).
43. Haines, M. E., Johnston, I. R., Mathias, A. P., and Ridge, D., *Biochem. J.* **115**, 881 (1969).
44. Hameister, H., and Richter, D., *Hoppe-Seyler's Z. Physiol. Chem.* **351**, 531 (1970).
45. Hasegawa, S., Fujimura, S., Shimizu, Y., and Sugimura, T., *Biochim. Biophys. Acta* **149**, 369 (1967).
46. Hayes, M. B., and Kaplan, N. O., *Fed. Proc.*, *Fed. Amer. Soc. Exp. Biol.* **31**, 452 (1972) (abstr.).
47. Hilz, H., and Kittler, M., *Hoppe-Seyler's Z. Physiol. Chem.* **349**, 1793 (1968).
48. Hilz, H., and Kittler, M., *Hoppe-Seyler's Z. Physiol. Chem.* **352**, 1693 (1971).
49. Honjo, T., *Taisha* **9**, 36 (1971).
50. Honjo, T., Nishizuka, Y., and Hayaishi, O., *Cold Spring Harbor Symp. Quant. Biol.* **34**, 603 (1969).
51. Honjo, T., Nishizuka, Y., Hayaishi, O., and Kato, I., *J. Biol. Chem.* **243**, 3553 (1968).
52. Honjo, T., Nishizuka, Y., Kato, I., and Hayaishi, O., *J. Biol. Chem.* **246**, 4251 (1971).
53. Honjo, T., Ueda, K., Tanabe, T., and Hayaishi, O., *in* "Metabolic Interconversion of Enzymes" (O. Wieland, E., Helmreich, and H. Holzer, eds.), p. 193. Springer-Verlag, Berlin and New York, 1972.
54. Johnston, I. R., Mathias, A. P., Pennington, F. A., and Ridge, D., *Biochem. J.* **109**, 127 (1968).
55. Johnston, I. R., Mathias, A. P., Pennington, F. A., and Ridge, D., *Nature (London)* **220**, 268 (1968).
56. Kalckar, H. M., *in* "The Mechanism of Enzyme Action" (W. D. McElroy

and B. Glass, eds.), p. 675. Johns Hopkins Press, Baltimore, Maryland, 1954.
57. Kandel, J., and Collier, R. J., Fed. Proc., Fed. Amer. Soc. Exp. Biol. 31, 871 (1972) (abstr.).
58. Kaplan, N. O., Colowick, S. P., and Nason, A. J., J. Biol. Chem. 191, 473 (1951).
59. Kato, I., Jap. J. Exp. Med. 32, 335 (1962).
60. Kato, I., Kagaku (Tokyo) 33, 15 (1963).
61. Kato, I., and Pappenheimer, A. M., Jr., J. Exp. Med. 112, 329 (1960).
62. Kloppstech, K., Steinbeck, R., and Klink, F., Hoppe-Seyler's Z. Physiol. Chem. 350, 1377 (1969).
63. Lipmann, F., and Tuttle, C., J. Biol. Chem. 159, 21 (1945).
64. Lucas-Lenard, J., and Lipmann, F., Annu. Rev. Biochem. 40, 409 (1971).
65. Marcus, A., and Gill, D. M., personal communication.
66. Marks, G. S., Marshall, R. D., and Neuberger, A., Biochem. J. 87, 274 (1963).
67. Matsubara, H., Hasegawa, S., Fujimura, S., Shima, T., and Sugimura, T., J. Biol. Chem. 245, 3606 (1970).
68. Matsubara, H., Hasegawa, S., Fujimura, S., Shima, T., Sugimura, T., and Futai, M., J. Biol. Chem. 245, 4317 (1970).
69. Michel, A., Zonen, J., Monier, C., Crispuls, C., and Dirkx, J., Biochim. Biophys. Acta 257, 249 (1972).
70. Miwa, M., and Sugimura, T., J. Biol. Chem. 246, 6362 (1971).
70a. Miyakawa, N., Ueda, K., and Hayaishi, O., Biochem. Biophys. Res. Commun. 49, 239 (1972).
71. Moehring, T. J., and Moehring, J. M., J. Bacteriol. 96, 61 (1968).
72. Moehring, T. J., Moehring, J. M., Kuchler, R. J., and Soloforofsky, M., J. Exp. Med. 109, 407 (1967).
73. Montanaro, L., and Sperti, S., Biochem. J. 105, 635 (1967).
74. Montanaro, L., Sperti, S., and Mattioli, A., Biochim. Biophys. Acta 238, 493 (1970).
75. Nagao, M., Yamada, M., Miwa, M., and Sugimura, T., Biochem. Biophys. Res. Commun. 48, 219 (1972).
76. Nakazawa, K., Ueda, K., Honjo, T., Yoshihara, K., Nishizuka, Y., and Hayaishi, O., Biochem. Biophys. Res. Commun. 32, 143 (1968).
77. Nashimoto, Y., Uchida, H., and Tsugita, K., Seikagaku 36, 693 (1964).
78. Nishizuka, Y., Ueda, K., and Hayaishi, O., in "Methods in Enzymology" (D. B. McCormick and L. D. Wright, eds.), Vol. 18B, p. 230. Academic Press, New York, 1971.
79. Nishizuka, Y., Ueda, K., Honjo, T., and Hayaishi, O., J. Biol. Chem. 243, 3765 (1968).
80. Nishizuka, Y., Ueda, K., Nakazawa, K., and Hayaishi, O., J. Biol. Chem. 242, 3164 (1967).
81. Nishizuka, Y., Ueda, K., Nakazawa, K., Reeder, R. H., Honjo, T., and Hayaishi, O., J. Vitaminol. 14, 143 (1968).

82. Nishizuka, Y., Ueda, K., Yoshihara, K., Yamamura, H., Takeda, M., and Hayaishi, O., *Cold Spring Harbor Symp. Quant. Biol.* **34**, 781 (1969).
83. Oikawa, A., Itai, Y., Okuyama, H., Hasegawa, S., and Sugimura, T., *Exp. Cell Res.* **57**, 154 (1969).
84. Okuyama, T., and Satake, K., *J. Biochem. (Tokyo)* **47**, 454 (1960).
85. Olivera, B. M., and Lehman, I. R., *Proc. Nat. Acad. Sci. U.S.* **57**, 1700 (1967).
86. Otake, H., Miwa, M., Fujimura, S., and Sugimura, T., *J. Biochem. (Tokyo)* **65**, 145 (1969).
87. Pappenheimer, A. M., Jr., and Brown, R., *J. Exp. Med.* **127**, 1073 (1968).
88. Preiss, J., Schlaeger, R., and Hilz, H., *FEBS Lett.* **19**, 244 (1971).
89. Raeburn, S., Collins, J. F., Moon, H. M., and Maxwell, E. S., *J. Biol. Chem.* **246**, 1041 (1971).
90. Raeburn, S., Goor, R. S., Collins, J. F., and Maxwell, E. S., *Biochim. Biophys. Acta* **199**, 294 (1970).
91. Raeburn, S., Goor, R. S., Schneider, J. A., and Maxwell, E. S., *Proc. Nat. Acad. Sci. U.S.* **61**, 1428 (1968).
92. Raynand, M., Bizzini, B., and Relyveld, E., *Bull. Soc. Chim. Biol.* **47**, 261 (1965).
93. Reeder, R. H., Ueda, K., Honjo, T., Nishizuka, Y., and Hayaishi, O., *J. Biol. Chem.* **242**, 3172 (1967).
94. Römer, V., Lambrecht, J., Kittler, M., and Hilz, H., *Hoppe-Seyler's Z. Physiol. Chem.* **349**, 109 (1968).
95. Roux, E., and Yersin, A., *Ann. Inst. Pasteur, Paris* **2**, 629 (1888).
96. Schneider, J. A., Raeburn, S., and Maxwell, E. S., *Biochem. Biophys. Res. Commun.* **33**, 177 (1968).
97. Shima, T., Fujimura, S., Hasegawa, S., Shimizu, Y., and Sugimura, T., *J. Biol. Chem.* **245**, 1327 (1970).
98. Shima, T., Hasegawa, S., Fujimura, S., Matsubara, H., and Sugimura, T., *J. Biol. Chem.* **244**, 6632 (1969).
99. Shimizu, Y., Hasegawa, S., Fujimura, S., and Sugimura, T., *Biochem. Biophys. Res. Commun.* **32**, 143 (1968).
100. Skogerson L., and Moldave, K., *Biochem. Biophys. Res. Commun.* **27**, 5 (1967).
101. Smulson, M. E., Henrikson, O., and Rideau, C., *Biochem. Biophys. Res. Commun.* **43**, 1266 (1971).
102. Smulson, M. E., and Rideau, C., *J. Biol. Chem.* **245**, 5350 (1970).
103. Smulson, M. E., Rideau, C., and Raeburn, S., *Biochim. Biophys. Acta* **244**, 269 (1970).
104. Sperti, S., and Montanaro, L., *Biochem. J.* **107**, 730 (1968).
105. Sperti, S., Montanaro, L., and Mattioli, A., *Chem. Biol. Interact.* **3**, 141 (1971).
106. Spiro, R. G., *Annu. Rev. Biochem.* **39**, 599 (1970).
107. Strauss, N., and Hendee, E. D., *J. Exp. Med.* **109**, 145 (1959).
108. Sugimura, T., *Progr. Nucl. Acid Res. Mol. Biol.* **13**, 127 (1973).

109. Sugimura, T., Fujimura, S., Hasegawa, S., and Kawamura, Y., *Biochim. Biophys. Acta* **138**, 438 (1967).

110. Sugimura, T., Fujimura, S., Hasegawa, S., Shimizu, Y., and Okuyama, H., *J. Vitaminol.* **14**, 135 (1968).

111. Sugimura, T., Yoshimura, H., Miwa, M., Nagai, H., and Nagao, M., *Arch. Biochem. Biophys.* **147**, 660 (1971).

112. Sung, S. C., and Williams, N. J., Jr., *J. Biol. Chem.* **197**, 175 (1952).

113. Traugh, J. A., and Collier, R. J., *Biochemistry* **10**, 2357 (1971).

114. Traugh, J. A., and Collier, R. J., *Biochem. Biophys. Res. Commun.* **40**, 1437 (1971).

115. Uchida, T., Gill, D. M., and Pappenheimer, A. M., Jr., *Nature (London), New Biol.* **233**, 8 (1971).

116. Uchida, T., Pappenheimer, A. M., Jr., and Harper, A. A., *Science* **175**, 901 (1972).

116a. Ueda, K., Miyakawa, N., and Hayaishi, O., *Hoppe-Seyler's Z. Physiol. Chem.* **353**, 844 (1972).

117. Ueda, K., Oka, J. Narumiya, S., Miyakawa, N., and Hayaishi, O., *Biochem. Biophys. Res. Commun.* **46**, 516 (1972).

118. Ueda, K., Reeder, R. H., Honjo, T., Nishizuka, Y., and Hayaishi, O., *Biochem. Biophys. Res. Commun.* **31**, 379 (1968).

119. Yamada, M., Miwa, M., and Sugimura, T., *Arch. Biochem. Biophys.* **146**, 579 (1971).

120. Yoshihara, K., *Biochem. Biophys. Res. Commun.* **47**, 119 (1972).

121. Zatman, L. J., Kaplan, N. O., and Colowick, S. P., *J. Biol. Chem.* **200**, 197 (1953).

122. Zimmerman, S. B., Little, J. W., Oshinsky, C. K., and Gellert, M., *Proc. Nat. Acad. Sci. U.S.* **57**, 1841 (1967).

Selected Topics on the Biochemistry of Spermatogenesis

IRVING B. FRITZ

Banting and Best Department of
Medical Research
University of Toronto
Toronto, Ontario, Canada

I. Introduction

A. General

Approximately 2300 years ago, Aristotle noted the analogy between the effects observed in castrated boys and those which follow the castration of immature birds (*142*). About 2200 years later, Berthold (1849) demonstrated that testis transplants were able to survive in castrated cockerels (capons), and that the noninnervated, transplanted testes secreted substances capable of allowing full de-

velopment of the characteristics of a sexually mature rooster
(wattles, combs, spurs, and aggressive sexual behavior). Berthold's
experiments provided the first convincing evidence for the presence
of agents later to be called hormones (8).

The two primary functions of the testis are to produce androgens
and to generate gametes. As might be expected from the length of
time during which the testis has been an active subject for inquiry,
the accumulated knowledge about testicular physiology is considerable. At present, information concerning the testis is being acquired
at an ever increasing rate, stimulated perhaps in part by a growing
interest in exploring all possible ways of controlling fertility. In
this climate, it is hardly surprising that large numbers of reviews
are appearing on various aspects of spermatogenesis. My excuse for
adding the present one rests partially upon the desire of a relative
novice to share enthusiasm generated while trying to collate material pertaining to various approaches to the investigation of the biochemistry of spermatogenesis. I shall not attempt the formidable
task of comprehensive coverage, but instead shall review primarily
those portions of the literature related to current personal research
interests in this area. In the remainder of this introductory section,
I shall endeavor to provide requisite background information concerning the biology of spermatogenesis, and to illustrate that spermatogenesis provides an interesting model for the investigation of
differentiation.

B. Fundamental Biological Background

Spermatogenesis may be defined as the process by which spermatozoa are generated from their progenitor cells. In all vertebrates
thus far investigated, the stem cells in the testis (type A_s spermatogonia)* give rise to differentiating spermatogonia which undergo
a series of subsequent mitoses. Eventually, a type B spermato-
gonium is formed, which is immediate precursor to the resting or

---

* The dynamics of stem cell renewal during spermatogenesis remain a
controversial subject. Differing viewpoints have recently been described in
detail (26, 66, 114). For purposes of the limited discussion in this review,
the abbrevation A_s represents the population of spermatogonial stem cells
in the testis which eventually generate type A_1 spermatogonia (66). Different
mechanisms may operate in various species (26).

preleptotene primary spermatocyte. After this final mitosis, a complex set of changes occurs within the diploid preleptotene spermatocyte, leading to the first reductive meiotic division. The prolonged prophase of this process is divided into four phases (leptotene, zygotene, pachytene, and diplotene), defined according to morphological characteristics of the nucleus. During the meiotic prophase, lasting for approximately 16 days in rats, homologous chromosomes are intimately paired. Synaptinemal complex formation then occurs, followed by the appearance of chiasmata, thereby facilitating the process of genetic recombination. After the completion of the first meiosis, two haploid secondary spermatocytes are produced, each containing the same amount of DNA as that present in somatic cells. The second meiotic division of both cells follows within 24 hours; it results in the generation of four haploid spermatids, each containing one-half the amount of DNA present in somatic cells. The spermatids then participate in a series of remarkable changes as they develop and eventually mature into spermatozoa, a process which requires approximately 18 days in rat testis.

Morphological changes associated with these complex processes are summarized in Fig. 1, taken from the publication by Dym and Clermont (40). Terminology employed in this review is the same as that used by Clermont (26) in describing the various cell types, stages of spermatogenesis, and cycles of the seminiferous epithelium. In the rat, each cycle is completed in 12–13 days, and four cycles are required for the production of mature spermatozoa from type A spermatogonia (26) (see cell association pattern in stage VIII of Fig. 1). The nature of the advancing front of cells during the cycles of the seminiferous epithelium can be visualized radioautographically in histological sections, as shown in Fig. 2, taken from the publication by Go et al. (53). One day after the in vivo administration of thymidine-^{3}H the germinal cells, labeled in their nuclear DNA, are preleptotene spermatocytes and spermatogonia (Fig. 2A). Twelve days after the injection of thymidine-^{3}H the most advanced germinal cells labeled are stage VIII pachytene spermatocytes (Fig. 2B). After an additional twelve days, the most advanced pulse-labeled germinal cells are seen to be spermatids (Fig. 2C). In the final cycle of the seminiferous epithelium, labeled spermatozoa are evident (Fig. 2D).

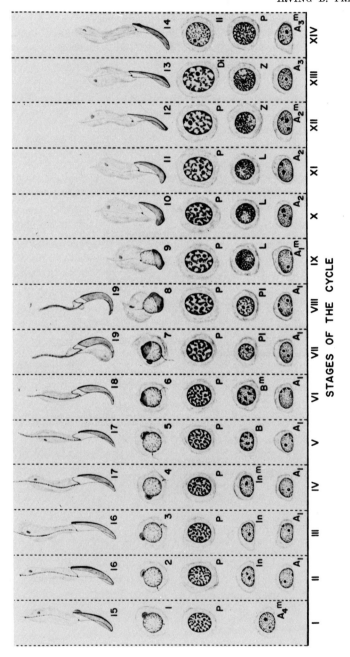

Spermatogenesis occurs exclusively within the seminiferous tubules of the testis. However, the tubule contains several types of cells other than germinal epithelial cells. For example, there are specialized peritubular cells in the several layers surrounding the tubule, and Sertoli cells are present in the basal layer. These cells play an integral role in tubular function and most probably constitute an important part in establishing the "blood–testis barrier." Other possible functions of the Sertoli cell will be discussed later. Consideration of the development of the mammalian testis (52), the nature of the testicular blood supply, lymphatic drainage, and tubular fluid (125), and the fine structure of various cells in the testis (17) have been extensively reviewed in the three-volume treatise, The Testis, edited by Johnson et al. (71).

The complexity of spermatogenesis (shown in Fig. 1) has perhaps tended to discourage biochemical investigations of this process. The heterogeneous populations of germinal cells within the seminiferous tubule, the existence of other tubular cells mentioned above, and the presence of androgen-synthesizing Leydig cells between the tubules, indicate some of the difficulties inherent in investigating the biochemistry of single cell types in the testis. In addition, a specialized tubular fluid bathes the germinal cells in a unique environment (128, 148). The existence of basal and adluminal compartments within the tubule, separated by specialized tight junctions between Sertoli cells (45), adds another level of complexity to biochemical investigations of spermatogenesis. However, recent

---

Fig. 1. Drawing from Dym and Clermont (40) illustrating the cellular composition of the 14 stages of the cycle of the seminiferous epithelium in the rat. Each column numbered with a Roman numeral shows the cell types present in one of the cellular associations found in cross sections of seminiferous tubules. Following cellular association (XIV), cellular association (I) reappears, so that the sequence starts over again. Other observations are identified by the following abbreviations: A_1, A_2, A_3, and A_4, four generations of type A spermatogonia; In, intermediate spermatogonia; B type B spermatogonia; Pl, preleptotene spermatocytes; L, leptotene spermatocytes; Z, zygotene spermatocytes; P, pachytene spermatocytes; Di, diakineses of primary spermatocytes; II, secondary spermatocytes. The arabic numerals designate spermatids during the various stages of maturation, defined according to changes in the acrosomic structure. Spermatogonia and spermatocytes undergoing mitosis are identified by the letter "m." Published with permission.

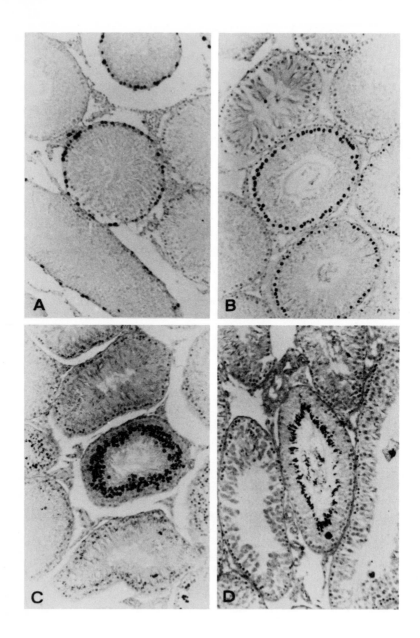

attempts to separate by physical means the various cell types in the testis have met with some degree of success (see Section II). In addition, it has been possible to eliminate some cell types by employing testes from sexually immature animals, or from animals which had previously been hypophysectomized or otherwise treated to alter testicular functions. For example, irradiation of the testis preferentially destroys spermatogonia (40), and placing testes within the abdomen (experimental cryptorchidism) eventually results in the loss of all advanced germinal cells (145, 148). Cryptorchidism is discussed in Section IV.

Spermatogenesis provides an interesting model for the investigation of developmental processes. The generation of specialized haploid cells, the spermatozoa, from the relatively less differentiated diploid spermatogonial stem cells is an obviously fascinating problem. Some of the changes which take place include the following. During their maturation, spermatids develop an acrosome which covers the anterior part of the head. The acrosome is rich in lysosomal enzymes (3), and these hydrolytic enzymes contained within the acrosome presumably function to facilitate penetration of the mature spermatozoan into the egg during fertilization. The nucleus of a spermatozoa almost fills the cell body or head. The condensed chromatin in the nucleus is tightly packaged with protamines or histones, the composition of the nucleoprotein varying among spermatozoa from different species (34). Spermatozoa contain scant cytoplasm and have very low levels of RNA. Accordingly, the density is very great (1.3), and the cell volume relatively small.

FIG. 2. Radioautographs of testicular sections at various times after the *in vivo* administration of thymidine-^{3}H to normal adult rats. All sections illustrated are from stage VIII of the seminiferous epithelial cycle. Initial magnification: ×100. (A) Tubules from a rat killed 1 day after thymidine-^{3}H injection. The labeled cells are preleptotene spermatocytes and spermatogonia. (B) Tubules from a rat killed 12 days after thymidine-^{3}H injection. The most advanced labeled cells are pachytene spermatocytes. (C) Tubules from a rat killed 24 days after having received thymidine-^{3}H. The most advanced labeled cells are mid-stage spermatids. (D) Microphotograph of a section of tubules from a rat killed 36 days after thymidine-^{3}H administration. The most advanced labeled cells are spermatozoa. From Go *et al* (53). Reproduced by permission of the National Research Council of Canada from *Can. J. Biochem.* **49**, 753–760 (1971).

A specialized array of mitochondria in mature spermatozoa appears in the highly organized mid-piece. The overall biochemical properties of the structurally altered mitochondria are, however, apparently the same as those of mitochondria from other tissues (56, 103). The mitochondria in the mid-piece are intimately related to fibers in the tail portion, where a characteristic 9 plus 2 array of filaments exists, embedded within a complex flagellum. It is this structure which provides a capability of motility to spermatozoa. The nature of the changes in structure of mitochondria during the development of spermatids has been thoroughly investigated (5, 43, 44).

During spermatogenesis, the profile of enzymes in the various cells is greatly altered. Some enzymes are deleted during development, while others are added (see Section III). Yet, mature spermatozoa contain enzymatic pathways which permit the utilization of a wide variety of substrates, including carbohydrates and nearly all classes of lipids. However, many of the biosynthetic enzyme systems are lost or greatly diminished, and rates of protein, DNA, or RNA synthesis in mature spermatozoa are negligible. An exhaustive review of some of the other properties of spermatozoa is provided by Mann (89). In a description of changes that occur during formation of mature spermatozoa, it is difficult to avoid teleological phrases. Spermatozoa appear to be relatively self-contained, self-propelled specialized cells which are being prepared to permit the fusion of the nucleus of a spermatozoan with that of an ovum during fertilization. The many structural and biochemical changes during development occur as though they were geared toward facilitating the achievement of this mission.

The generation of spermatozoa from spermatogonia would seem to provide a clear example of differentiation. However, it should be emphasized that the stem cells originally in the testis are themselves already differentiated. The activities of these relatively less differentiated spermatogonia (type A_s) are restricted to the generation of daughter cells like themselves, or to the formation of type A_1 spermatogonia. Once the type A_1 spermatogonia are formed, the process appears to be biologically irreversible, and the cell must then either continue its progression of programmed divisions, leading to increasingly more differentiated cells, or die (66–68). The precise conditions under which the testicular stem cells are initially formed and transported to the embryonic testis are unknown. The possible role of the supporting cells in influencing germinal cells during the embryonic development of the testis have been considered by Jost (72), and the migration of gonocytes into the testis has been reviewed by Peters (115).

II. Techniques for Obtaining Enriched Populations of Various Cell Types from the Testis

A. General Procedures

The easiest way to obtain relatively homogeneous populations of germinal epithelial cells is to use a source already enriched in the cell type desired. During development in all species, germinal cells in tubules from the immature testis are primarily spermatogonia. At a later stage of maturation, spermatocytes are formed, but spermatids are absent. In the mature testis the predominant cell types are spermatids and spermatozoa. The enrichment of particular classes of germinal cells in testes from animals at different stages of development can be used to great advantage in the isolation of various germinal cells from testes of salmonid fishes. Spawning salmon have relatively huge testes (about 6% of the body weight), containing primarily advanced spermatids and spermatozoa. Testes in this state can be obtained from salmonids during their natural spawning season, or from fishes made to develop by injections of pituitary hormones (91). Friedrich Miescher in the 1870's used this approach in his isolation of purified preparations of "nuclein," later to be called nucleic acids by his students. Miescher prepared the nucleic acids from sperm heads of cells obtained from testes of spawning Rhine salmon. An interesting account of these researches has been written by Tepperman (141).

In rats, similar experimental approaches can be taken, since testes from 14-day-old animals contain primarily spermatogonia, whereas spermatids do not appear in testes until about day 26 to 30 of life (29). Functional changes in testes from adult rats can be evoked by various manipulations. As mentioned above, irradiation of adult rat testes results in the preferential destruction of spermatogonia (40). Irradiation of pregnant rats during the latter stages of gestation alters testicular development in male offspring (60). Many seminiferous tubules in neonatal rats which had been irradiated in this fashion contain primarily Sertoli cells, while germinal epithelial cells are virtually absent. Another example of altered cell profiles in the testis is seen in hypophysectomized animals. It has been well appreciated since the observations of Smith (*129, 130*) that removal of the pituitary is followed by a diminution in size of the testes. By 30 days post-hypophysectomy in rats, the predominant cell types remaining in the tubule are spermatogonia, Sertoli cells, and a reduced number of spermatocytes (*28*). A similar pattern is obtained in cryptorchid testes. The loss of all advanced germinal cells occurs in rats within 2–3 weeks after testes are placed within the abdominal cavity (*145*).

Several types of mutant mice have been characterized which have disturbances in testicular development. For example, the Sxr (sex reversal) mouse is a genetic female which appears as a phenotypic male. The testis of the adult Sxr mouse contains only Sertoli cells within the tubule, although Leydig cells are present in abundance between tubules (*22*). The Tfm (testicular feminization) mouse is a genetic male which appears as a phenotypic female, having a rudimentary testis within the abdomen. The seminiferous tubule in this mutant contains no germinal cells more advanced than spermatocytes (*85*), but this block may represent a nonspecific secondary effect of cryptorchidism. Additional types of mice exist (translocations T16H, T145, and T199) which have spermatogenesis arrested at various other stages of development (*84a*).

In testes which structurally contain peritubular lymphatic sinusoids (*45*), it appears feasible to tease seminiferous tubules from testes, leaving the intertubular tissue and Leydig cells behind. This has been successfully performed by Christensen and Mason (*24*) in the cases of tubules from testes of guinea pigs, rats, hamsters, rabbits, dogs, and normal mice, but not in the cases of testes from cats or men. Isolated tubules can be relatively easily obtained in fairly long segments from intact, hypophysectomized, cryptorchid, and irradiated rats, thereby allowing preliminary separation of various populations of seminiferous tubular cells from Leydig cells.

Once the tubules are obtained, the cells within the tubules can be freed by mechanical or by enzymatic treatment. Separation of testicular cells from tubules or whole testis into various populations

has been partially achieved by use of the "Staput Procedure" (*75*).
Miller and Phillips (*101*) had previously shown that cells of different sizes from mouse spleen could be separated according to their
different sedimentation velocities in a specially designed chamber,
the "Staput," kept at unit gravity. The cells are initially layered
onto a shallow albumin gradient, and then allowed to sediment at
4°C for 4 hours. Lam *et al.* (*75*) applied this technique to the separation of cells obtained from mouse testis, the tubules of which had
been mechanically fragmented with a razor blade assembly. These
workers obtained six peaks of cells in the different Staput fractions,
and showed that the slowest moving peak (0.75 mm/hour) consisted
of spermatozoa. A much faster moving fraction of cells (10
mm/hour) was reported to consist primarily of diplotene and late
pachytene cells (*75*).

In an extension of these studies, Meistrich (*97*) improved procedures by preparing cell suspensions from mouse testis with a trypsin-DNase treatment in a manner which did not release nuclei from
spermatogonia, and which digested damaged cells. Subsequent dilution with serum albumin and slow cooling to 5°C resulted in high
yields of cells which excluded trypan blue, and which appeared to
retain integrity when measured by additional criteria. With this
treatment, larger percentages of intact spermatogonia and resting
spermatocytes were obtained than with the mechanical blades assembly procedure. Trypsin treatment had also been shown to be
useful for the preparation of cell suspensions from rat testis during
early stages of development (*146*). It is of interest that the mechanical preparations with the razor blade assembly damaged spermatogonia and preleptotene spermatocytes more than did the trypsin procedures, but the former treatment exerted less damage to
spermatids (*97*). It is apparent that the manner of preparing the
cell suspension is of considerable importance in determining the
various populations of cells obtained in different fractions after employing differential sedimentation velocity separation in the Staput
chamber. Because mechanical preparations tend to release some late
spermatids in conjunction with Sertoli fragments (*98*), as well as
to release nuclei from spermatogonia and preleptotene spermatocytes, it appears that these preparations are less generally useful
than those obtained after trypsin treatment. However, the late-stage

primary spermatocytes (pachytene and diplotene) are fortunately resistant to the mechanical blade assembly procedure, and remain intact. Therefore this procedure can still be satisfactorily employed in kinetic experiments in which the progression of preleptotene spermatocytes to the more rapidly sedimented pachytene spermatocytes is being investigated.

Germinal epithelial cells at the same stage of development within the seminiferous tubule are joined by cytoplasmic bridges (16, 46, 102, 111). These bridges are presumably broken during the separation of cells by either the mechanical or trypsin procedures. Sealing of at least some of the broken bridges occurs, as shown in Fig. 3. The site of the former bridge is easily identified because of the characteristic electron dense structure on the plasma membrane which is present in all cytoplasmic bridges between germinal epithelial cells. However, some cells may remain joined, and certainly multinucleate cells may be formed when the tubules are treated with trypsin. This is most pronounced for the case of spermatids (98), adding still another complicating factor to the interpretation of Staput profiles.

The relative numbers of various populations of germinal epithelial cells have been determined in different fractions obtained after separation of testicular cell suspensions by sedimentation velocity fractionation in the Staput chamber. In the case of rat testicular cells, the suspensions were prepared with the mechanical blade assembly procedure (53), whereas with mouse testicular cells, the suspensions were prepared with the trypsin-DNase procedure (98). The distribution of rat testicular cells, identified microscopically, in the various peaks is shown in Fig. 4. In these experiments, each nucleus was counted as a single cell. As can be seen from these data, no single fraction contained a homogeneous population of cells. However, fractions around number 120 (corresponding to a sedimentation velocity of 1.25 mm/hour) consisted primarily of elongated spermatids and spermatozoa; fractions 60–80 (corresponding to sedimentation velocities of approximately 6–8 mm/hour) contained mixtures of round spermatids and leptotene plus zygotene spermatocytes and fraction 40 (corresponding to a sedimentation velocity of 11.2 mm/hour) was enriched in pachytene plus diplotene spermatocytes. However, even this rapidly sedimenting fraction was contaminated with large numbers of spermatids,

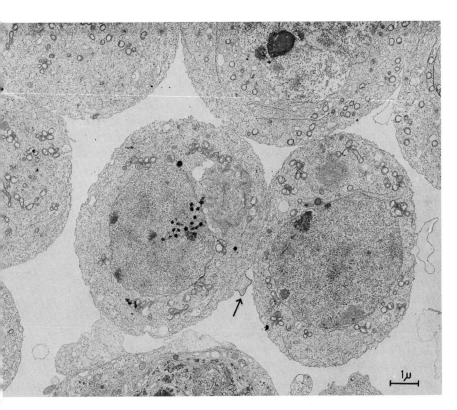

Fig. 3. Electron microphotograph of section of pachytene spermatocytes. Cells were obtained from the testis of a 26-day-old normal rat. The cell suspension was prepared by trypsinization of teased testicular tissue, and cells were layered in the Staput for fractionation by the sedimentation velocity technique (75, 146). Fractions sedimenting at 10–12 mm/hour were pooled, centrifuged, and fixed in buffered glutaraldehyde. They were subsequently postfixed in buffered osmium tetroxide, dehydrated and then embedded in Epon and Araldite. Sections were cut on a Porter-Blum MT-II microtome with a diamond knife, and then stained with uranyl acetate and lead hydroxide. Am RCA EM-3G with Kodak plate positive film was used to obtain the microphotograph. Arrow points to the site of a resealed intercellular bridge. Unpublished observations of V. L. W. Go, R. G. Vernon, E. Whitter, and I. B. Fritz. ×7000.

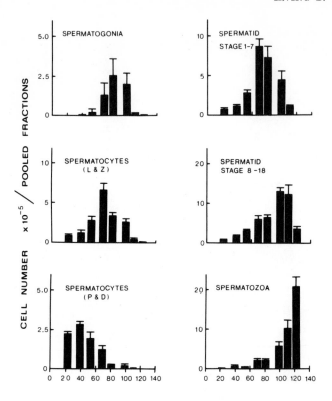

FRACTION NUMBER

Fig. 4. The absolute number of various cell groups of pooled fractions obtained from four separate Staput runs of cell suspensions from adult normal rat testes (mean ± SEM). The number of cells of each class are plotted as a function of the fraction number, as described by Go *et al.* (*53*). The abbreviations and nomenclature used are the same as those defined in the legend to Fig. 1. From Go *et al.* (*53*). Reproduced by permission of the National Research Council of Canada from *Can. J. Biochem.* **49**, 753–760 (1971).

many of which were multinucleate (*53*). These populations could be made less heterogeneous by employing cell suspensions from testes obtained from 26-day-old rats, in which case fraction 40 contained 75% primary spermatocytes and only 12% spermatids (*146*). Staput fractionation was not employed to separate cells from testes of 7- or 14-day-old rats, since spermatogonia comprised over 90% of the total cell suspensions (146).

Meistrich et al. (98) obtained more homogeneous populations of cells in fractions from trypsin preparations of mouse testis. In fractions sedimenting at 9.5 mm/hour, pachytene spermatocytes comprised 73% of the total cells, with multinucleate cells comprising much of the remainder. In fractions sedimenting at 4.1 mm/hour, round spermatids comprised 77% of the total while elongated spermatids accounted for an additional 10%. In other fractions, the separation was less complete (98). The presence of multinucleate cells and fragments of elongated spermatids in various fractions, and the existence of heterogeneous populations of cells in nearly all Staput fractions, obviously limits the usefulness of these procedures for biochemical investigations requiring isolation of single cell types. It is therefore apparent that fundamental technical problems remain to be solved in relation to the optimal preparation of suspensions of mononucleate cells from testis. As the technology for obtaining undamaged cell suspensions improves, the Staput fractionation procedure can be expected to permit better resolution of various cell populations. Even in their present form, however, the procedures described have considerably facilitated investigations of the kinetics of differentiation of germinal epithelial cells in mice (75, 97) and rats (53). Other examples of applications of these procedures, in which the differences in sedimentation velocities of different classes of germinal cells have been exploited in pulse-label experiments to follow the advancing front of developing cells, are given in Section II, B.

Nuclei isolated from germinal epithelial cells may also be separated and classified according to their microscopic structure. Nuclei prepared from hamster testis were separated by density gradient centrifugation (143, 144). With procedures employed, there was partial purification of pachytene cell nuclei from spermatid nuclei. Better separation was achieved with the Staput procedure (99). Elongated spermatid preparations from mouse testis, prepared with Staput fractionation of trypsin-treated suspensions, were subjected to cationic detergent exposure in order to free nuclei from the cells (83). These nuclei were then analyzed by equilibrium density centrifugation in a linear colloidal silica gradient. The buoyant density of nuclei increased during steps 11–14 of spermiogenesis, thereby allowing separation of homogeneous populations of nuclei from various classes of spermatids (83). Stages 1 to 9 spermatids contained nuclei having a density of 1.067 gm/ml, whereas the density increased to 1.073, 1.095, 1.112, and 1.170 gm/ml in nuclei from stages 10, 11, 12, and 13 spermatids, respectively. In stage 14 spermatids, the density increased to 1.2 gm/ml. Morphological characteristics of the nuclei separated were established with electron microscopic examinations (83). Findings summarized provide a methodology for the investigation of the biochemical changes occurring in nuclei during spermatogenesis. While the procedures outlined preclude the examination of cytoplasmic processes in various classes of germinal cells, the availability of relatively homogeneous populations of apparently undamaged nuclei should considerably facilitate the study of nuclear events during differentiation and development of spermatids.

B. Examples of Application of Cell Separation Procedures In Investigations of Spermatogenesis

Studies on differentiation kinetics of testicular germinal epithelial cells of the mouse (75, 97) and the rat (53) have been performed with the aid of the Staput fractionation procedure. A pulse of thymidine-³H, injected in vivo, permits incorporation of the base into DNA being synthesized at that time by spermatogonia and preleptotene spermatocytes. Most of the DNA is preserved as these cells develop, undergo meiosis, and eventually generate spermatids which progress to form spermatozoa (Fig. 2) The differing sedimentation velocities of these cells permits investigation of the differentiation process. The appearance of labeled DNA in various cell fractions in the Staput profiles corresponds to information obtained previously with the aid of radioautographic examinations by Monesi (104), and various cytologic investigations by others (26, 76, 113). In spite of the obvious heterogeneity of cell populations in most of the various fractions (Fig. 4), the Staput profiles allow identification of the labeled cell type because of the great differences in sizes, and therefore in sedimentation velocities, among preleptotene spermatocytes, pachytene spermatocytes, and spermatids. The progress of the advancing front of developing labeled cell types is thereby readily ascertained, even though the total populations of cells in fractions containing the DNA-labeled cells lacks homogeneity. Examples of Staput profiles of labeled testicular cells are shown in Fig. 5, taken from the publication of Go *et al.* (*54*). In cell suspen-

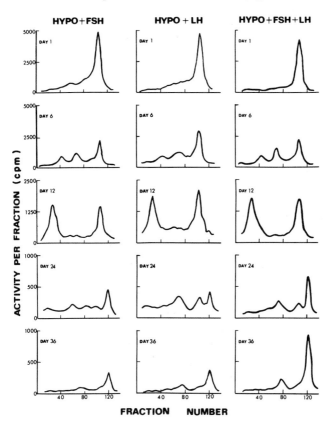

FIG. 5. Staput profiles of the distribution of labeled DNA in cell fractions from rat testes. The number listed in the upper left-hand corner of each panel represents the number of days after the administration of thymidine-^{3}H before testicular cell suspensions were prepared. All injected rats had been hypophysectomized (HYPO) 30 days previously, and then were treated with either follicle-stimulating hormone (FSH), luteinizing hormone (LH), or both (FSH + LH) for the periods shown. Counts per minute per fraction are plotted against the fraction number, the origin being fraction 135 in each case (*53*, *54*). For details, the original, publication should be consulted (*54*). Reproduced by permission of the National Research Council of Canada from *Can. J. Biochem.* **49**, 768–775 (1971).

sions from testes of rats killed 1 day after the administration of thymidine-³H, most of the label was in nuclei from preleptotene spermatocytes and spermatogonia. In hypophysectomized rats receiving adequate therapy (follicle-stimulating hormones, or FSH, and luteinizing hormone, or LH), the labeled cells in testes obtained 6 days after injection of thymidine-³H included leptotene–zygotene and early pachytene spermatocytes (see last column of Fig. 5). Twelve days after administration of the labeled DNA precursor, approximately half the counts were found in the pachytene spermatocytes, while the remaining half of the label was in nuclei from spermatogonia and preleptotene spermatocytes. This distribution would be expected, since at the time of the pulse label, half of the germinal epithelial cells in the DNA-synthetic phase would have been preleptotene spermatocytes, and the other half would be comprised of all classes of spermatogonia (104). At 24 days after thymidine-³H administration, the labeled cells included round and elongated spermatids, whereas at 36 days after thymidine-³H administration, the bulk of the label was in elongated or mature spermatids (see bottom right-hand profile in Fig. 5). These results on the advancing fronts of labeled cells were comparable to those observed in intact adult rats (53, 54). In hypophysectomized rats which had not received adequate hormonal therapy, the Staput profiles indicated an absence of adequate progression of spermatogenesis beyond the stage of pachytene spermatocytes (first two columns of profiles in Fig. 5). Data obtained from this sort of approach (53, 54, 75, 97) are in good general agreement with observations and conclusions made earlier by others, based primarily on classic cytologic and quantitative radioautographic investigations (for reviews, see 26, 104, 121, 133). Results shown on Staput profiles in Fig. 5 should be compared with radioautographs illustrated in Fig. 2. In differentiation kinetic experiments, in which Staput profiles are obtained of testes previously pulse-labeled with suitable precursors, it therefore appears possible to obtain the equivalent of "instant radioautography." The disadvantages imposed by the presence of heterogeneous populations of cells in Staput fractions (Fig. 4) may in large part be overcome by following the advancing front of pulse-labeled cells. These procedures in their present form are not as precise as those of quantitative radioautography, and it is therefore apparent that conclusions drawn must be validated with the more laborious and time-consuming microscopic techniques. However, information obtained in experiments such as those shown in Fig. 5 may often be more useful than that which can be obtained with qualitative radioautography, such as that illustrated in Fig. 2, where some of the tubules appear to be totally unlabeled.

An interesting application of the cell separation techniques involved studies on the site of synthesis of protamine by mouse testicular germinal cells (74). Lam and Bruce (74) demonstrated that protamine is synthesized by spermatids which have progressed approximately 19 days beyond the S phase which occurs in preleptotene primary spermatocytes. This was confirmed by demonstrating that mouse spermatozoa, isolated from the vasa deferentia, were maximally labeled with arginine-³H 12 days after the *in vivo* administration of the pulse, whereas they were maximally labeled with thymidine-³H approximately 31–32 days after the *in vivo* administration of the DNA precursor. In the first case, nearly all the labeled arginine was in protamines from spermatozoa which had been observed to be spermatids shortly after the arginine injection. In the latter case, the labeled thymidine was in DNA from spermatozoa which had been observed to be preleptotene spermatocytes shortly after thymidine administration, as determined by sedimentation velocity analysis (74). It is of interest that Monesi (105) had shown earlier with radioautographic techniques that labeled amino acids were incorporated into those mouse spermatids which were past stage 11 and up to stage 16. The observations of Lam and Bruce (74) may therefore be regarded as a confirmation and extension of the report of Monesi (105). The agreement between the two investigations serves as an indirect validation of the Staput procedures. These and other related applications of the Staput fractionation technique have recently been discussed by Bruce and Meistrich (15).

III. Enzyme Profile Changes During Spermatogenesis

A. General

It is assumed that increased knowledge concerning the alterations in enzyme patterns in advancing germinal epithelial cells will prove advantageous in understanding differentiation during spermatogenesis (9). At the very least, this information should provide biochemical data which can be correlated with the morphological changes shown in Fig. 1, and it should also indicate suitable enzyme markers for following developmental changes. In addition, the findings may give insights into aspects of the biochemistry of differentiation and development. Examples of changes in enzyme activities in various classes of germinal epithelial cells will be described within these contexts.

A review by Bishop (9) of testicular enzymes as fingerprints in the study of spermatogenesis briefly covered the cytochemical demonstrations of germinal epithelial enzymes, and dealt specifically with the nature of changes in testis-specific hexokinase and the X-band of lactate dehydrogenase during spermatogenesis in various species. In addition, testicular sorbitol dehydrogenase was reviewed in considerable detail with respect to comparative distribution; localization in advanced germinal epithelial cells but not in spermatogonia; and the characterization of the purified enzyme (9).

Other reviews concerned with enzyme patterns in various testicular germinal cells include these by Fox and Fox (50) and Blackshaw (10). The composition of spermatozoa and the metabolism by isolated spermatozoa have been intensively examined. More investigations have been performed on spermatozoa than on any other germinal cell, most probably because spermatozoa are the only homogeneous class of germinal epithelial cells which thus far can be obtained in large yields. Earlier results of investigations in this area have been reviewed by Salisbury and Lodge (122) and by Mann (89). Mohri et al. (103) have presented a very good analysis of the enzymatic and metabolic properties of isolated mitochondria from spermatozoa. More recently, Peterson and Freund (116) have surveyed glycolytic enzyme activities in spermatozoa. A detailed summary of the metabolic properties of spermatozoa is beyond the scope of this review.

With the exception of investigations on spermatozoa, most of the reports concerned with determination of activities of various enzymes in germinal epithelial cells have relied upon analyses of whole testis preparations from animals during different stages of development. This approach affords useful information about the testis, and often permits deductions about general metabolic activities and enzyme formation in various classes of germinal cells. However, it does not permit conclusions concerning the enzyme content of specific populations of germinal cells. In efforts to circumvent this difficulty, histochemical analyses have been attempted (*9, 10, 50, 117, 118*). These procedures, under optimal conditions, permit localization of enzymes to specific cells. When coupled with radioautographic techniques, they allow identification of cell types associated with synthesis of DNA, RNA, and proteins (*104, 105*). However, it is very difficult, if not impossible, from these observations to quantitate the amount of enzymatic activity per cell. It would be ideal if homogeneous cell populations could be obtained from the testis for direct chemical analyses, or for isolation of pure enzymes from single cell types so that antibodies could be subsequently prepared against the specific cell proteins. This objective remains to be achieved, but a beginning has been attempted. A few examples follow.

B. Changes in Enzyme Activities during the Development of Advanced Spermatocytes and Early Spermatids

1. LACTATE DEHYDROGENASE

Spermatogonia, the least differentiated of the germinal epithelial cells, have low levels of several enzymes which appear in larger amounts in developing spermatocytes. For example, the species of lactate dehydrogenase (EC 1.1.1.27) designated as LDH-X is a unique isozyme thus far found only in testis. It first appears in the developing testis of mice at approximately 14–15 days postnatally (*11, 55*) and in rats at approximately 20 days (*12*). It is at this time that pachytene spermatocytes are being formed in the seminiferous tubules. However, in Wistar rats, LDH-X was reported to appear initially in testis from 35-day-old rats (*151*), a time at which immature spermatids are present in abundance. This isozyme is apparently localized in the mitochondrial fraction of testis homogenates and in the mid-piece of spermatozoa (*25*). Subsequent work has indicated that the "heavy mitochondrial fractions" of rat testis, presumably derived primarily from spermatids and the mid-piece of spermatozoa, contained most of the specific LDH-X (*37*).

The cellular subfraction which is rich in LDH-X contains mitochondria of characteristic morphology, which first appear in advancing spermatocytes (36).

The testicular isozyme contains subunits different from those of skeletal muscle (type A) or heart (type B), and have been designated type C (153). LDH-X catalyzes the oxidation of DL-α-hydroxybutyrate and DL-α-hydroxyvalerate to the corresponding keto acids, whereas the usual forms of lactate dehydrogenase are specific for lactate (2). These differences in substrate specificity have been employed to demonstrate that LDH-X in testis is localized in tubular cells and not in interstitial (Leydig) cells (2, 11, 55, 153). In hypophysectomized adult mice, the LDH-X content of testis decreased linearly as a function of time after hypophysectomy (11). The same phenomenon has been reported in hypophysectomized rats (151). The combined data have been interpreted to indicate that LDH-X is most probably synthesized during the time when preleptotene spermatocytes are undergoing changes leading to the formation of pachytene spermatocyes (12, 55), and that these changes are under the control of both FSH and androgens (11, 151). Properties and kinetic characteristics of LDH-X from guinea pig testis have also recently been described (6). Earlier observations on LDH-X in testes from mice and rats have been confirmed and extended with LDH-X prepared from this species. LDH-X appears to be a homopolymer of type C subunits, whereas a different isozyme, LDH-X', acts as a hybrid of A and C subunits (6). The different classes of germinal cells thus seem to possess specificity with respect to the type of LDH subunit synthesized. Apparently, spermatogonia synthesize only A and B types; developing spermatocytes and spermatids synthesize C types which can hybridize with A to form LDH-X; and advanced spermatids and spermatozoa preferentially contain homopolymers of C type LDH subunits which form the unique LDH-X' in guinea pig testis (6).

The functions of these different isozymes of LDH are not clear, but the potential usefulness of the unique LDH-X subunit as a marker during different stages of spermatogenesis is obvious.

2. CARNITINE ACETYLTRANSFERASE

Marquis and Fritz (90) reported that carnitine acetyltransferase (EC 2.3.1.7) (CAT) levels in testis from 14-day-old rats were very low relative to those present in testis from 35-day-old or adult animals. This observation suggested the possibility that spermatogonia, which are the predominant germinal epithelial cells present in testes from 14-day-old rats, were deficient in CAT, whereas spermatocytes and spermatids had high levels. This interpretation was reinforced by the observations that CAT activities were greatly reduced in regressed cryptorchid testes, and that isolated spermatozoa had high CAT levels (90). In a follow-up to these experiments, Vernon *et al.* (146) demonstrated that testicular CAT levels first began to rise in developing rats at a time corresponding to the period at which mature pachytene and diplotene primary spermatocytes, and immature spermatids, were first being generated in abundance, namely at 24-32 days after birth. In addition, testicular CAT levels in hypophysectomized rats fell to less than 10% of intact adult values by approximately 30-35 days after the removal of the pituitary (146). At this time, the percentages of the different germinal cells become constant and consist primarily of spermatogonia and spermatocytes (28). The observations cited were consistent with the interpretation that CAT levels are increased during the transitions involved in the formation of advanced spermatocytes and spermatids, and that CAT activity in spermatogonia is low.

In efforts to validate these conclusions more directly, CAT levels were determined in different classes of germinal epithelial cells isollated from testes of 7-, 14-, and 26-day-old rats, as well as from testes of adult animals. Testicular cells were fractionated in the Staput sedimentation chamber, and CAT activities were determined in cells obtained from various fractions enriched in particular populations of germinal cells. Results obtained indicated that fractions rich in spermatogonia had approximately 0.1 unit/10^8 cells, whereas fractions enriched in leptotene-zygotene, pachytene and diplotene cells had approximately 1.8, 4.5, and 6.2 CAT units/10^8 cells. Enzyme levels in elongated spermatids and spermatozoa were reduced to approximately 2.2 units/10^8 cells (146). It was therefore suggested that CAT is probably synthesized during the time of primary spermatocyte development, and that the CAT formed remains in spermatids generated from the spermatocytes. From available data, the possibility of continuing formation of CAT by spermatids could not be ruled out (146). Independent of this possible interpretation, it is apparent that CAT provides a convenient enzyme marker with which to follow the transition from spermatogonia to spermatocytes in suitable systems.

The functions of CAT in germinal epithelial cells are not known. Cellular CAT activities cannot be correlated with rates of long-chain fatty acid oxidation, since spermatogonia-rich testicular populations have higher rates of palmitate oxidation than do testicular cell suspensions enriched in advanced spermatocytes or spermatids, even though CAT levels are lowest in spermatogonia (79).

In testes from *Drosophila melanogaster*, CAT levels have been reported to be highest in spermatids (51). This conclusion was based on determination of enzyme activities in cells obtained from three consecutively sliced segments of the testis. According to the information given, spermatocytes were present only in the "upper middle" region of the fly testis, and in this fraction CAT levels were low (51). If these sections were indeed rich in advanced spermatocytes, it appears that an interesting species difference exists in which CAT is synthesized during spermatocyte development in rats, but only during spermatid development in flies. In flies which were sterile as a consequence of a choline-deficient diet, testicular CAT levels were low, whereas CAT activities in other tissues examined in these flies were within normal values. It therefore appears that formation of CAT is intimately associated with the development of normal advanced germinal cells (51).

3. ADENYL CYCLASE AND CYCLIC NUCLEOTIDE PHOSPHODIESTERASE

Levels of testicular adenyl cyclase (AC) progressively increased during development of rats from 25 days to 60 days of age (62). This is the time of sexual maturity during which advanced spermatocytes and spermatids are formed. The pattern of change in AC activity during this interval generally resembles that described above for the changes in testicular CAT levels in rats (146). Highest specific activities of AC in subcellular fractions were observed to be in the mitochondrial fraction (62).

A unique type of cyclic nucleotide phosphodiesterase "f" (PDE) begins to appear in testis at 40 days of age, achieving adult levels in testis from 50-day-old rats (107). During this period, AC activity increases in a parallel fashion to 5-fold the levels present at 20 days (62). However, PDE activity seems to be associated primarily with the formation of spermatids, while the initial increase in the activity coincides with the time at which advanced primary spermatocytes are being generated. In testes from immature rats (20-day-old or younger), only the "c" type of PDE is present. Both forms of PDE are located primarily in the high speed supernatant subcellular fractions (107).

No reports have yet appeared showing which isolated testicular cells contain these enzymes. It may be deduced that the "c" type of PDE is located in spermatogonia, Leydig cells, and/or Sertoli cells present in testes from immature animals. It may be inferred that AC must be present in interstitial cells, since LH increases cAMP formation, and cAMP increases androgen synthesis in these cells (31, 120). The site of action of LH on androgen formation by testis has been reviewed by Hall (58, 59) and by Eik-Nes (41).

The probable location of adenyl cyclase in various classes of testicular cells can also be inferred from information available concerning the effects of gonadotropic hormones on AC activities in testis preparations. Both FSH and LH have been reported to increase the amounts of cAMP formed by homogenates of dog or rat testis (109). In addition, Kuehl et al. (73) assayed AC activities in testis slices and in collagenase digests of testis in the presence and in the absence of gonadotropic hormones. The collagenase digests were thought to consist primarily of seminiferous tubules, but it appears likely that these preparations may have been contaminated by varying amounts of Leydig cells (38). FSH incubation in vitro with various preparations of testis obtained from immature or from hypophysectomized rats resulted in stimulation of AC activity in a manner which was not duplicated by incubation with low concentrations of LH. However, LH addition also increased AC activities in testis preparations from normal adult rats, whereas FSH did not (73).

Using preparations of isolated seminiferous tubules, Dorrington et al. (39) demonstrated that LH did not influence adenyl cyclase activity. In these isolated tubules, however, AC activity was greatly increased by FSH. In contrast, in Leydig cell-rich preparations, LH enhanced AC activity but FSH did not (38, 39). By employing tubular preparations from regressed cryptorchid or hypophysectomized rats and from irradiated rats, it was possible to eliminate

various populations of germinal epithelial cells. FSH acted to increase AC activity in tubules from all groups of rats listed. From
these and other data presented, Dorrington and Fritz (*38*) concluded that the increased cAMP levels in seminiferous tubules incubated with FSH resulted primarily from hormonal stimulation of
Sertoli cells. In these studies, there was no evidence that any germinal epithelial cells directly responded to FSH with an increase in
AC activity (*38*). These conclusions are consistent with observations of Murphy (*110*), showing an FSH effect on the morphology
of Sertoli cells and with the demonstration by Castro *et al.* (*20,
21*) that FSH preferentially binds to Sertoli cells.

From combined observations cited, the interpretation emerges
that FSH increases adenyl cyclase activity in the Sertoli cell of
the seminiferous tubule but not in the Leydig cell, whereas LH increases AC activity in the Leydig cell but not in the tubule. If,
however, the Sertoli cell and the Leydig cell were the only testicular
cells containing adenyl cyclase, it would be difficult to account for
the changes in AC activity in developing rat testes reported by
Hollinger (*62*). By the time AC activity begins to increase most
rapidly in testis (40–60 days), the Leydig cells and Sertoli cells are
already well formed. Hollinger's data are most easily interpreted
by assuming that adenyl cyclase appears in advanced spermatocytes
and in spermatids, and that these cells do not necessarily respond
directly to gonadotropic stimulation with an increased AC activity.
The observation that cAMP addition increased phosphorylation of
histones and protamines in spermatids (*69*) supports the suggestion
that a functional adenyl cyclase may be present in the advanced
germinal cells.

The properties of a solubilized adenyl cyclase preparation from
the testis of chinook salmon have been characterized by Menon and
Smith (*100*). Just as in the above studies with rat testis, AC activity
in salmon testis slices was observed to be increased by addition of
salmon pituitary gonadotropin (*100*).

The kinetic properties of type "f" PDE from rat testis have been
compared with those of the nonspecific type "c" isozyme, and it
has been ascertained that the K'_m for cAMP of the unique testicular
isozyme is 24-fold lower than the corresponding K'_m of the type
"c" PDE (*107*). As Monn *et al.* (*107*) indicated, the appearance of a

specific isozyme of PDE in advanced germinal cells of the testis is reminiscent of the appearance of LDH-X in spermatocytes and spermatids (see Section III, B, 1).

The possible functions of the adenyl cyclases in each of the classes of testicular cells remain to be determined. It would appear quite likely that cAMP formation in the Leydig cell, in response to LH stimulation, is associated with androgen formation; it would appear possible that increased cAMP formation in the Sertoli cell, in response to FSH stimulation, is associated with important aspects of Sertoli cell function; but the possible functions of the adenyl cyclases in advanced spermatocytes and in spermatids are totally speculative. They may be associated with the regulation of histone and protamine phosphorylation in spermatids (*69, 91*).

4. Enzymes Primarily Associated with Meiosis

The processes occurring during the prolonged prophase in the primary spermatocyte permit genetic recombination to take place prior to reductive division. The mechanisms involved in relation to chromosome pairing and chiasma formation during the meiotic prophase remain almost totally mysterious. However, certain unique enzymes must operate to make these chromosomal rearrangements possible. For example, specific endonucleases, polynucleotide ligases and polynucleotide kinases might be expected to be present during meiosis which are not necessarily present during mitosis. In addition, specific enzymatic machinery must exist in order to synthesize and organize the synaptinemal complex (*30, 108*). Thus far, very little biochemical work has been reported which helps to resolve these questions, and almost all the information available is based on cytologic and cytogenetic observations. This lack of knowledge may be attributed in part to the fact that proper conditions have not yet been established in which mammalian spermatocytes can be cultured synchronously *in vitro*, resulting in an orderly completion of meiosis.

However, Howell and Stern (*65*) have been able to examine some of the biochemical changes occurring during meiosis *in vitro* in *Lilium*. They have assayed the activities of a specific endonuclease, polynucleotide kinase, and polynucleotide ligase present in microsporocytes synchronously cultured *in vitro*. The specific activities of each of these enzymes was altered during meiotic prophase in a characteristic manner, with all three showing highest activities during the pachytene stages. In all three cases, enzyme activities rapidly fell shortly after the completion of the diplotene stage. A specific meiotic endonuclease was absent from the cell during all stages of meiosis except during those of zygotene and pachytene (65). The properties of each of the relevant enzymes and reactions were described. It was concluded that the data were consistent with the concept that the highest frequency of recombination occurs during the later zygotene and early pachytene stage of meiosis, a time at which the breakage and reunion of DNA is catalyzed by enzymes active in meiosis (65).

The existence of a unique "meiotic protein" was first demonstrated in meiotic cells of *Lilium* (63). It is localized in nuclei of cells which are in meiotic prophase. The meiotic protein is absent from somatic cells, and from germinal cells not in meiotic prophase. This protein binds preferentially to single-stranded DNA and has *in vitro* properties very similar to those of gene 32 of bacteriophage T4 (63). It was suggested that the "meiotic protein" may be involved in the process of genetic recombination, perhaps during the intimate pairing of homologous chromosomes (63).

In a recent study, the same "meiotic protein" was identified in extracts of nuclei prepared from advanced spermatocytes obtained from rat, bull, and human testes (64). From available information, Hotta and Stern (64) concluded that nuclei from these cells in meiotic prophase contained a DNA-binding protein having the same properties as those of gene 32 of phage T4 and of "meiotic protein" from *Lilium*. The broad phylogenetic distribution demonstrated implies that this protein may be of considerable importance in genetic recombination during meiosis. Further developments in this field will obviously be followed with great interest. When more precise localization of the purified protein can be achieved, an increased understanding of its functional significance during meiosis can be anticipated.

Other information on the biochemistry of meiosis, and summaries of results from investigation of gametogenesis in lilies, may be obtained from the reviews of Stern and Hotta (136, 137) and that of Monesi (106).

C. Changes in Enzyme Activities during the Development of Late Spermatids

1. Enzymes Associated with Protamine Synthesis, and Modification of Histones and Protamines in Testis

The investigations of Lam and Bruce (74) demonstrated that protamine synthesis in mouse testis occurs only in certain classes of spermatids. In experiments on spermatogenesis in trout testis, Dixon and colleagues have shown that protamines are synthesized during the terminal stages of spermatogenesis, and that these protamines replace histones previously present. Concomitantly, nucleohistones are transformed into nucleoprotamines of spermatids, and chromosomal template activity is greatly reduced [for review, see Marushige and Dixon (91)].

The enzymes associated with protamine synthesis are present in the cytoplasm of trout spermatids. After the protamines are synthesized on disome complexes, they then enter the spermatid nucleus (80, 82). Three different protamines have been isolated, each of which is synthesized at a characteristic time during testis development (81). The factors responsible for the removal of repression of genes responsible for coding the synthesis of protamines in spermatids are unknown. However, this phenomenon apparently does not occur in any other cell type. The appearance of the large disome peak in spermatids has been used as a marker for the presence of messenger RNA specific for protamine synthesis in the ribosomal fraction of spermatids (80, 82). It is not clear, however, whether this mRNA is synthesized during an earlier stage and stored until protamines are programmed to be synthesized, or whether the specific mRNA is synthesized immediately prior to or concomitant with disome formation during the late stages of spermatogenesis. In mouse testis, most RNA synthesis is thought to take place prior to the late stages of spermatid formation (105).

By the time protamine synthesis is initiated, histone synthesis in trout testis ceases. The apparent subsequent sequence of events include a phosphorylation of the seryl residues of the protamines, which then enter the nucleus, where they bind to DNA (92). The nature of the phosphorylation process, and the properties of the

cAMP-dependent protein kinase which catalyzes the phosphorylation, have been described (69, 70). Before the phosphorylated
protamines are bound to the nuclear DNA, the preexisting nuclear
histones are modified by phosphorylation of seryl residues and by
acetylation of lysyl residues (19, 138). It is proposed that the modified histones then become detached from the DNA, and the phosphorylated protamines replace them. Subsequently, the protamines
bound to DNA are dephosphorylated, forming nucleoprotamines
characteristic of mature sperm. The modified histones leave the nucleus, where they are degraded (91).

The enzymatic changes described result in chemical transformation of the proteins associated with DNA in the chromatin of developing germinal epithelial cells. The template activity of native
nucleoprotamine is inactive in support of RNA synthesis, presumably because the DNA is very tightly packaged by the protamine,
which appears to behave as a nonspecific repressor molecule (91).

2. HYALURONIDASE

Evidence documenting the increased activities of hyaluronidase,
α-glycerophosphate dehydrogenase, and sorbitol dehydrogenase during late spermiogenesis has been reviewed by others (9, 50).
Hyaluronidase is located in the acrosomes of spermatids and spermatozoa (4, 61, 88, 131, 132). In the rat, testicular hyaluronidase
begins to appear at approximately 33 days postnatally, and there-
after increases some 400-fold, reaching its peak in 60-day-old animals (86).

This coincides with the beginning development of the proacro-
somal cap of spermatids, and most rapid rises occur during the
maturation of spermatids. Hypophysectomy of rats 28 days old prevented the subsequent increase in testicular hyaluronidase activity
which otherwise occurs in normal rats. Prolonged continuous maintenance therapy of hypophysectomized rats, with FSH plus LH or
testosterone, restored enzyme levels to normal testicular values
(86). However, in rats deprived of gonadotropins for 2–3 days after
hypophysectomy, normal enzyme levels were not achieved until
30–35 days after the beginning of therapy (86). In adult rats, hypophysectomy was followed by decreased levels of testicular
hyaluronidase, coinciding with the loss of spermatids (134). In

agreement with findings previously cited concerning the acrosomal location of hyaluronidase, highest specific activities of hyaluronidase were observed in testicular subcellular fractions sedimenting with the lysosome-rich particles (86).

3. SUMMARY

From combined data, it may be concluded that the appearance of hyaluronidase is a marker enzyme for the initial formation of spermatids. The presence of protamine synthesizing enzymes marks a later stage of spermatid maturation. In contrast, the formation of advancing primary spermatocytes is accompanied by the appearance of LDH-X and carnitine acetyltransferase. The unique phosphodiesterase "f" may also provide a marker for early spermatids, but the precise cellular location of this isozyme remains to be determined. Adenyl cyclase is present in Leydig cells and Sertoli cells, as well as in advancing germinal epithelial cells. The possibility that isozymes of adenyl cyclase may also exist has not been examined thus far because of the difficulty of purifying this membrane-bound enzyme.

It is evident that only a beginning has been made in characterizing the various germinal cells biochemically. Nevertheless, marker proteins have been identified which allow one to follow the generation of specific classes of germinal cells, and some promise of insight is available concerning the role of enzymes apparently associated with the genetic recombination process occurring during the zygotene and pachytene stages of meiosis. Most impressively, considerable information has been accumulated concerning the processes involved in the transformation of nucleohistones to nucleoprotamines during the latter stages of spermiogenesis.

IV. Biochemical Changes during Cryptorchidism

A. General

In most mammals, a scrotum exists into which the testes descend during development. If the testes stay within the abdomen, spermatogenesis does not proceed. This condition (cryptorchidism) can easily be induced experimentally in rats by placing testes into the abdominal cavity, after which all germinal cells except spermatogonia and preleptotene spermatocytes degenerate within 2 weeks. Leydig cells and Sertoli cells remain relatively normal during this period, as determined morphologically, and by functional criteria used to assess evidence of continuing testosterone secretion by the Leydig cells. The relevant literature demonstrating the temporal sequence of various structural changes occurring in the seminiferous tubule of cryptorchid animals has been reviewed by Waites (*147*), Waites and Setchell (*148*), and Van Denmark and Free (*145*).

The fundamental impairment of spermatogenesis at elevated testicular temperatures is unknown. From an evolutionary viewpoint, it is not clear what advantages were bestowed in mammals by having the germinal epithelium develop at intratesticular temperatures approximately 4–5°C below the body core temperature of mammals. Nonscrotum bearing homeothermic vertebrates (birds) obviously have no difficulty in producing spermatozoa in an environment maintained at 37–40°C. Yet, in most mammals, the lower temperature of 32–33°C is permissive for spermatogenesis to proceed, but an elevation to 37°C blocks one or more sites of the developmental stages. From experiments with local heating of the testes of rats or rams, it has been concluded on the basis of histological examination that primary spermatocytes undergoing meiosis are the most temperature-sensitive germinal cells (*148*). Spermatocytes in other stages of the cell cycle, and immature spermatids, are also damaged by exposure to normal body temperature. In contrast, mature spermatozoa are obviously capable of survival at 37° in the epididymis, vas deferns, and vagina for the length of time required for fertilization. The underlying mechanisms which could account for the temperature sensitivity of some, but not all, germinal epithelial cells remain to be discovered.

The cessation of spermatogenesis which occurs during cryptorchidism can be reversed by lowering the intratesticular temperatures, provided sufficient germinal epithelial stem cells remain in the tubule. Biologically, this is most evident in seasonal breeders, such as antelopes or bats, in which the testes have lesser amounts of spermatozoa formation during those seasons of the year in which the female is not in heat (*33*).

Chowdhury and Steinberger (*23*) have recently reported experiments in which testes from newborn rats were transplanted into either scrotal or cryptorchid testes of adult animals. From results obtained, these authors concluded that "a cryptorchid milieu does not interfere with the initiation of spermatogenesis, but prevents its progression and maintenance" (23). Moreover, if the cryptorchid host testis were inserted into the scrotum, spermatogenesis in the transplanted testis from neonatal rats proceeded identically initially, and then went to completion. There were, however, only patchy areas of regeneration in the host testis, which had previously been in the abdomen for 30 days prior to reinsertion into the scrotum. The fact that the neonatal testis, while in the cryptorchid milieu, allowed the initial development to pachytene primarily spermatocytes was interpreted to indicate that the initiation of spermatogenesis was different from the maintenance of spermatogenesis in adult animals. The appearance of advanced spermatocytes in the transplanted testis, under conditions in which the host cryptorchid testis had no germinal cells advanced this far, does demonstrate a difference in the two processes. If the transplanted neonatal testis were kept in the cryptorchid milieu, degeneration eventually occurred to the same extent as in the host (23).

Attempts have been made to correlate the *in vivo* impairments in cryptorchid testes with metabolic changes observed in isolated testis slices incubated at various temperatures. The incorporation of labeled lysine into proteins by testicular slices from rats, mice, and hamsters was lower at 37°C than at 32°C. However, in testicular slices from rabbit, guinea pig, and dog, the incorporation was higher at 37°C (18). In rats, the inhibitory effects of incubation at elevated temperature on protein synthesis were attributed primarily to effects on spermatids (35). In all the above species, however, cryptorchidism results in impairments of spermatogenesis *in vivo*. It would therefore appear that the diminution in protein synthesis observed in rat spermatids incubated at higher temperatures is a secondary effect which cannot be invoked to account for the *in vivo* failure of spermatogenesis at 37°C.

Recently, Nishimune and Komatsu (112) examined the effects of temperature on rates of metabolic pathways in seminiferous tubules from mouse testis. These tubules were incubated in the presence of labeled substrates for 2 hours at various temperatures, following a prolonged preincubation period. The incorporation of uridine-^{3}H into RNA was greatly increased in tubules maintained at 38°C, as opposed to those incubated at 30°C. The incorporation of serine-^{14}C into protein was slightly less at 38°C then at 30°C. In contrast, the incorporation of thymidine-^{3}H into DNA at 38°C was half that observed at 32°C (112). This inhibition was also observed in isolated germinal epithelial cells, but there was no decrease in apparent DNA synthesis by isolated testicular supporting cells incubated at the higher temperature. These observations were interpreted to indicate that those cells capable of synthesizing DNA (the preleptotene spermatocytes and/or spermatogonia) were the most temperature-sensitive (112). These findings are somewhat difficult to correlate with previous demonstrations that existing pachytene spermatocytes and spermatids readily degenerate when exposed to elevated temperatures (148). If the impairments in DNA synthesis by germinal cells incubated at 38°C are the "primary and essential cause of the testicular atrophy" during cryptorchidism (112), the damage to existing cells which do not synthesize DNA appears difficult to rationalize. Evidence for the possible involvement of an impairment in phosphofructokinase activity during the early stages of cryptorchidism has also been presented (42). Combined results suggest that several classes of germinal epithelial cells are sensitive to temperature elevation, and that this sensitivity is reflected by metabolic impairments in major biosynthetic and metabolic pathways at more than one site in spermatogenesis. Pachytene spermatocytes appear to be most readily affected, and immature spermatids only slightly less so. However, preleptotene spermatocytes, and perhaps other germinal epithelial cells including spermatogonia, may be adversely influenced by sustained exposure to a temperature appreciably higher than that normally prevailing in the intratesticular environment in the scrotum (145, 148).

B. Possible Involvement of Lysosomes in Cells of Cryptorchid Testes

1. EVIDENCE OF INCREASED THERMAL LIABILITY OF LYSOSOMES FROM CERTAIN CLASSES OF GERMINAL CELLS

Lee and Fritz (77) recently examined the hypothesis that one of the factors which renders certain classes of germinal epithelial cells heat sensitive may reside in the nature of the properties of lysosomes from these cells. Lysosome-rich particles prepared from normal adult rat testes were incubated at various temperatures, and the release of lysosomal enzymes into the medium was measured. It was observed that these particles, when incubated at 37°C, rapidly released arylsulfatase, β-N-acetylhexosaminadase, and acid phosphatases. In contrast, lysosome-rich particles from liver incubated at 37°C at neutral pH had much lower rates of release of these enzymes. At temperatures of 25°C or lower, the rates of release of lysosomal enzymes from testicular particles were approximately the same as those from liver lysosomes (77). Interestingly, the lysosome-rich particles from pigeon testis displayed a temperature sensitivity which was equivalent to that of liver lysosomes. This may be correlated with the fact that in pigeons the testis is located normally within the abdomen. It therefore appeared that the lysosomes from certain germinal cells of adult rat testis were unique, with respect to the enhanced lability of enzyme release at 37°C. When cryptorchid testes were allowed to regress to the point that chiefly Sertoli cells and spermatogonia remained, lysosome-rich particles from these gonads behaved essentially identically to those prepared from liver. Similar results were obtained in lysosome-rich particles prepared from testes of regressed hypophysectomized rats (77), and from testes of 14-day-old rats (77a). From these experiments, it was concluded that spermatocytes and spermatids from rat testis contain lysosomes which, when incubated at 37°C, readily permit egress of lysosomal enzymes. In contrast, all other cell populations examined contain lysosomes which are more stable under these conditions of incubation (77). In other experiments, it was shown that lysosome-rich particles prepared from spermatozoa isolated from the vas deferens of rats did not display the increased temperature sensitivity characteristic of lysosome-rich particles from adult rat testis (77a).

In view of these findings, it was then considered useful to investigate the possibility that the degeneration of germinal epithelial cells in cryptorchid testes might be intimately associated with the release of lysosomal enzymes in vivo. Efforts were made to estimate the relative amounts of lysosomal enzymes liberated from lysosomes into the cytoplasm in vivo at varying times after inserting testes into the abdomen. The activities of β-N-acetylhexosaminidase and arylsulfatase in testicular cytosol fractions were observed to increase 2- to 5-fold, respectively, within 40 hours after the operation (77). In contrast, the decreases in testicular weight and protein content, which are known to follow the production of cryptorchidism, were not discernible until 120–140 hours postoperatively. These data were compatible with the interpretation that certain germinal cells of rat testis contain lysosomes which are temperature sensitive and that these lysosomes may release their enzymes into the cytosol shortly after the onset of cryptorchidism. From these data, it is not possible, however, to determine whether the lysosomal changes preceded the events leading to cellular degeneration, or whether these abnormalities occur in response to a more fundamental common cause. For example, it is possible that several types of membranes synthesized by spermatocytes have properties that result in alteration of permeability characteristics when exposed to an environmental temperature of 37°C. The observed differences in lysosomes from advancing germinal epithelial cells may therefore be simply a reflection of this general type of change. However, it appears reasonable to suggest that the increased release of hydrolytic enzymes at 37°C from lysosomes having these unique properties may be intimately associated with the cessation of spermatogenesis which occurs in cryptorchid testes (77). Blackshaw and Hamilton (13) examined the histological and histochemical changes occurring in rat testes after they were immersed into a water bath at 42° or 43° for 30–60 minutes. Alterations observed were interpreted to indicate that the lysosomes may have been changed in pachytene spermatocytes in response to injury of various cellular membranes at critical stages of development (13).

2. GENERAL PROPERTIES OF TESTICULAR LYSOSOMES

Allison and Hartree (3), who examined the lysosomal enzymes in the acrosomes of spermatozoa from several species, have concluded that "the acrosome is a specialized lysosome which evolved to facilitate fertilization in multicellular organisms." In acrosomes from ram spermatozoa, it was shown that the following lysosomal enzymes were present: acid phosphatase, β-N-acetylglucosaminidase, phospholipase A, protease, and probably arylsulfatase (3).

Recently, arylsulfatase has been clearly localized in acrosomes of rat spermatids (*123*). In other studies previously cited (see Section III, C, 2), neuraminidase and hyaluronidase have also been found in acrosomes (*131, 132*). Cytological and histochemical results reviewed by Allison and Hartree (*3*) are consistent with the concept that the developing acrosome contains enzymes which are thought to be characteristic of lysosomes. Some enzymes [i.e., hyaluronidase (*86*) and the esterase-hydrolyzing boromochloroindoxyl acetate (*3*)] are present only in the proacrosomes of developing spermatids and the acrosomes of spermatozoa. This implies that the lysosomes in spermatogonia may contain a population of enzymes different from those in the acrosome of the developing spermatids.

Indirect support for this possibility is derived from the observation that testicular β-glucuronidase activity is apparently present only in Sertoli and spermatogonial cells (*87*). As testes mature, the specific activity of this enzyme, which is high in lysosomal fractions from testes of immature rats, steadily declines during normal development. If, however, rats are hypophysectomized on day 28 of life, the specific activity of testicular β-glucuronidase no longer falls but instead is increased to levels previously found in testes from 10- to 15-day-old animals (*87*). In these regressed hypophysectomized rats, treatment with FSH plus LH or testosterone stimulated spermatid formation. Concomitantly, there was a fall in the specific activity of testicular β-glucuronidase. Data described were interpreted to signify that the specific activity of β-glucuronidase is highest when the seminiferous epithelium is comprised primarily of spermatogonia. The interpretation is consistent with histochemical observations that β-glucuronidase appeared to be restricted to spermatogonia. Histochemical evidence of activity could not be detected in primary spermatocytes or spermatids (*87*). On the other hand, hyaluronidase is not present in spermatogonia, but its activity is high in spermatids (*86*).

It is therefore conceivable that the population of lysosomes in spermatogonia which contains β-glucuronidase is lost during development, and different populations of lysosomes exist in spermatocytes and spermatids. This supposition, if valid, is consistent with the normal thermal lability characteristics of lysosomes from cells present in regressed or in immature testes, as opposed to the increased temperature sensitivity of populations of lysosomes from advanced germinal cells (77).

Szego and colleagues (139, 140) have advanced the viewpoint that steroid hormones act in a specific manner to labilize the membranes of lysosomes isolated from target tissues. Of special interest in relation to topics discussed above, testosterone treatment *in vivo* was observed to enhance the labilization of membranes from lysosomes of preputial glands but not from those of uterus (140). The relevance of these interesting observations to possible influences of testosterone on spermatogenesis remains to be examined. It appears plausible to consider the possibility that lysosomal changes during spermatogenesis could contribute to developmental processes, and could be intimately associated with the impairments observed in cryptorchid testes.

V. Hormonal Control of Spermatogenesis

Steinberger (133) has recently discussed this topic in detail, and has reviewed the difficulties involved in attempting to determine the manner in which FSH and testosterone are required to permit spermatogenesis to proceed. Ever since Smith (129, 130) discovered that hypophysectomy of rats leads to an arrest of spermatogenesis, endocrinologists have been attempting to sort out the precise nature of the hormonal dependency. Aside from the "obvious" requirement for LH to permit androgen synthesis by the Leydig cell (41, 58, 59, 120), and the generalized requirement for testosterone at several stages of spermatogenesis (1, 7, 32, 57, 135), confusion persists. Until very recently, it was not unequivocally established that FSH played an essential role in spermatogenesis. Even now, some of the investigations reviewed by Steinberger (133) challenge the viewpoint that FSH is required for normal tubular function (14).

Administration of FSH to young rats was followed by increased rates of protein, phospholipid, and RNA synthesis by testis preparations from these animals (93–96, 152). These effects were not obtained when FSH was added *in vitro* to isolated testis slices. Addition of FSH to testis preparations *in vitro* increased adenyl cyclase activity (39, 73, 109). As previously stated, it appears likely that FSH actions on the Sertoli cell account for its stimulation of adenyl cyclase in isolated seminiferous tubules (38). This conclusion is consistent with demonstrations that labeled FSH is bound to Sertoli cells (*21, 23*). Earlier quantitative histological work by Murphy (*110*) had indicated that FSH treatment alters the morphology of Sertoli cells in regressed hypophysectomized rats, without influencing Leydig cells. However, a subsequent report by von Berswordt-Wallrabe *et al.* (*7*) failed to confirm that FSH necessarily induces a secretory hypertrophy of the Sertoli cells, evaluated by qualitative histological examination. It is clearly established, however, that hypophysectomy results in alterations in the structure of Sertoli cells, which regress to a form reminiscent of those in the immature testis (*26, 28*). The precise role of FSH in Sertoli cell metabolism remains to be established. It is an important point to resolve, because the probable functions of the Sertoli cell in spermatogenesis are multiple.

For example, the Sertoli cells appear to be responsible for maintaining the blood-testis barrier (*45, 125, 128, 148*). Sertoli cells have been implicated in this role in large part because of their unique morphology and because of the nature of the tight junctions between Sertoli cells (*45, 48, 111*). In addition, functional changes in the blood-testis barrier may be correlated with changes in Sertoli cell structure. For example, the barrier appears in immature rats at a time when the Sertoli cells are forming intercellular contacts (*47, 125–128*). In addition, the Sertoli cells have been implicated in the origin of seminiferous tubular fluid (*124, 128*). The nature of the tubular fluid may be crucial in establishing the optimal milieu in which spermatogenesis proceeds, and it is not inconceivable that hormones indirectly influence germinal epithelial cell development by virtue of their effects on cells controlling the blood-testis barrier and the formation of tubular fluid (*45, 125*).

The mode of action of testosterone on spermatogenesis remains obscure. Because of the complexities of this system and the absence of a suitable *in vitro* assay, the actions of testosterone on spermatogenesis have been less well explored than the actions of testosterone on other processes. Wilson (*150*) has recently reviewed possible mechanisms involved in testosterone and dihydrotestosterone actions on various target glands. Other recent reviews have been concerned with androgen actions on prostate and seminal vesicle (*78, 149*). The formation of dihydrotestosterone from testosterone by testicular tissue has been demonstrated by Folman *et al.* (*49*) and Rivarola *et al.* (*119*). It is therefore possible that dihydrotestosterone is the hormone active in influencing tubular function, just as it is in other tissues (*150*).

It is painfully apparent that all too little is known about the biochemical aspects of testosterone and FSH actions on spermatogenesis, beyond the fact that they are both required to restore spermatogenesis in regressed hypophysectomized rats (*54, 84, 133*). However, available information suggests many leads concerning the general sites to explore. The problem would be made considerably more amenable to approach if it were possible to culture germinal epithelial cells or seminiferous tubules under defined conditions so that desired stages of spermatogenesis could be made to proceed *in vitro*. Unfortunately, only limited success has thus far been achieved with organ culture of rat testis, and germinal epithelial cells have not progressed beyond pachytene spermatocytes *in vitro* (*133, 135*). The prospects of using hormones as triggers with which to examine the control of different aspects of spermatogenesis will undoubtedly continue to stimulate investigations on the mode of hormone actions on these processes.

In spite of its considerable complexity, spermatogenesis continues to remain one of the more attractive systems to investigate as a model for differentiation.

ACKNOWLEDGMENTS

Work reviewed which emanated from this laboratory was supported by grants from the Canadian Medical Research Council and the Banting Research Foundation. It is a pleasure to express my gratitude to several colleagues for stimulating discussion of topics reviewed. I especially wish to thank Drs. W. R. Bruce, Leo Lee, and Jenny Dorrington for a critical reading of the manuscript. I am grateful to Drs. M. L. Meistrich, W. R. Bruce, and Y. Clermont for allowing me access to articles prior to publication. The excellent secretarial assistance of Ms. Barbara Smith and Judy Birchall in assembling this manuscript is happily acknowledged.

REFERENCES

1. Albert, A., *in* "Sex and Internal Secretions" (W. C. Young, ed.), 3rd ed., Vol. 1, pp. 305–365. Williams & Wilkins, Baltimore, Maryland, 1961.
2. Allen, J. M., *Ann. N.Y. Acad. Sci.* **94**, 937 (1961).

3. Allison, A. C., and Hartree, E. F., *J. Reprod. Fert.* **21**, 501 (1970).

4. Allison, A. C., and Young, M. R., *Life Sci.* **3**, 1407 (1964).

5. André, J., *J. Ultrastruct. Res., Suppl.* **3**, 1–185 (1962).

6. Battellino, L. J., and Blanco, A., *Biochim. Biophys. Acta* **212**, 205 (1970).

7. Berswordt-Wallrabe, R. von, Steinbeck, H., and Neumann, F., *Endokrinologie* **53**, 35–42 (1968).

8. Berthold, A. A., in "Great Experiments in Biology" (M. L. Gabriel and S. Fogel, eds.), p. 58–60. Prentice-Hall, Englewood Cliffs, New Jersey, 1955.

9. Bishop, D. W., in "Reproduction and Sexual Behavior" (M. Diamond, ed.), p. 261. Indiana Univ. Press, Bloomington, 1969.

10. Blackshaw, A. W., in "The Testis" (A. D. Johnson, W. R. Gomes, and N. L. Van Demark, eds.), Vol. 2, pp. 73–123. Academic Press, New York, 1970.

11. Blackshaw, A. W., and Elkington, J. S. H., *Biol. Reprod.* **2**, 268 (1970).

12. Blackshaw, A. W., and Elkington, J. S. H., *J. Reprod. Fert.* **22**, 69 (1970).

13. Blackshaw, A. W., and Hamilton, D., *J. Reprod. Fert.* **24**, 151 (1971).

14. Boccabella, A. V., *Endocrinology* **72**, 787 (1963).

15. Bruce, W. R., and Meistrich, M. L., in "Cell Differentiation" (R. Harris, T. Allin, and D. Viza, eds.), pp. 295–299. Munksgaard, Copenhagen, 1972.

16. Burgos, M. H., and Fawcett, D. W., *J. Biophys. Biochem. Cytol.* **1**, 287 (1955).

17. Burgos, M. H., Vitale-Calpe, R., and Aoki, A. in "The Testis" (A. D. Johnson, W. R. Gomes, and N. L. Van Demark, eds.), Vol. 1, pp. 551–649. Academic Press, New York, 1970.

18. Buyer, R., and Davis, J. R., *Comp. Biochem. Physiol.* **17**, 151 (1966).

19. Candido, E. P. M., and Dixon, G. H., *J. Biol. Chem.* **246**, 3182 (1971).

20. Castro, A. E., Alonso, A., and Mancini, R. E., *J. Endocrinol.* **52**, 129 (1972).

21. Castro, A. E., Seiguer, A. C., and Mancini, R. E., *Proc. Soc. Exp. Biol. Med.* **133**, 582 (1970).

22. Cattanach, B. M., Pollard, C. E., and Hawkes, S. G., *Cytogenetics* **10**, 318 (1971).

23. Chowdhury, A. K., and Steinberger, E., *J. Reprod. Fert.* **29**, 173 (1972).

24. Christensen, A. K., and Mason, N. R., *Endocrinology* **76**, 646 (1965).

25. Clausen, J., *Biochem. J.* **222**, 207 (1969).

26. Clermont, Y., *Physiol. Rev.* **52**, 198 (1972).

27. Clermont, Y., and Harvey, S. C., *Ciba Found. Colloq. Endocrinol.* [*Proc.*] **16**, 173 (1967).

28. Clermont, Y., and Morgantaler, H., *Endocrinology* **57**, 369 (1955).

29. Clermont, Y., and Perey, B., *Amer. J. Anat.* **100**, 241 (1957).

30. Comings, D. E., and Okada, T. A., *Exp. Cell Res.* **65**, 104 (1971).

31. Cooke, B. A., Van Beurden, W. M. O., Rommerts, F. F. G., and van der Molen, H. J., *FEBS Lett.* **25**, 83 (1972).
32. Courot, M., Hochereau-de Reviers, M. T., and Ortavant, R., *in* "The Testis" (A. D. Johnson, W. R. Gomes, and N. L. Van Demark, eds.), Vol. 1, pp. 339. Academic Press, New York, 1970.
33. Cowles, R. B., *Quart. Rev. Biol.* **40**, 341 (1965).
34. Davis, J. R., and Langford, G. A., *in* "The Testis" (A. D. Johnson, W. R. Gomes, and N. L. Van Demark, eds.), Vol. 2, pp. 259–306. Academic Press, New York, 1970.
35. Davis, J. R., Morris, R. N., and Hollinger, M. A., *Amer. J. Physiol.* **207**, 50 (1964).
36. De Domenech, E. M., Domenech, C. E., Aoki, A., and Blanco, A., *Biol. Reprod.* **6**, 136 (1972).
37. De Domenech, E. M., Domenech, C. E., and Blanco, A., *Arch. Biochem. Biophys.* **141**, 147 (1970).
38. Dorrington, J. H., and Fritz, I. B., *Endocrinology* in press (1973).
39. Dorrington, J. H., Vernon, R. G., and Fritz, I. B., *Biochem. Biophys. Res. Commun.* **46**, 1523 (1972).
40. Dym, M., and Clermont, Y., *Amer. J. Anat.* **128**, 265 (1970).
41. Eik-Nes, K. B., *in* "The Androgens of the Testis" (K. B. Eik-Nes, ed.), pp. 1–47. Dekker, New York, 1970.
42. Ewing, L. L., and Schanbacher, L. M., *Endocrinology* **87**, 129 (1970).
43. Favard, P., and André, J., *in* "Comparative Spermatology" (B. Baccetti, ed.), pp. 415–429. Academic Press, New York, 1970.
44. Fawcett, D. W., *Biol. Reprod.* **2**, Suppl. 2, 90 (1970).
45. Fawcett, D. W., Leak, L. V., and Heidger, P. M., *J. Reprod. Fert., Suppl.*, **10**, 105 (1970).
46. Fawcett, D. W., Susumu, I., and Slautterback, D., *J. Biophys. Biochem. Cytol.* **5**, 453 (1959).
47. Flickinger, C. J., *Z. Zellforsch. Mikrosk. Anat.* **78**, 92 (1967).
48. Flickinger, C. J., and Fawcett, D. W., *Anat. Rec.* **158**, 207 (1967).
49. Folman, Y., Haltmeyer, G. C., and Eik-Nes, K. B., *Amer. J. Physiol.* **222**, 653 (1972).
50. Fox, B. W., and Fox, M., *Pharmacol. Rev.* **19**, 21 (1967).
51. Geer, B. W., and Newburgh, R. W., *J. Biol. Chem.* **245**, 71 (1970).
52. Gier, H. T., and Marion, G. D., *in* "The Testis" (A. D. Johnson, W. R. Gomes, and N. L. Van Demark, eds.), Vol. 1, pp. 1–45. Academic Press, New York, 1970.
53. Go, V. L. W., Vernon, R. G., and Fritz, I. B., *Can. J. Biochem.* **49**, 753 (1971).
54. Go, V. L. W., Vernon, R. G., and Fritz, I. B., *Can. J. Biochem.* **49**, 768 (1971).
55. Goldberg, E., and Hawtrey, C., *J. Exp. Zool.* **164**, 309 (1967).
56. Gonse, P. H., *in* "Spermatozoan Motility," Publ. No. 72, pp. 99–132. Amer. Ass. Advan. Sci., Washington, D.C., 1962.

57. Greep, R. O., *in* "Sex and Internal Secretions" (W. C. Young, ed.), 3rd ed., Vol. 1, pp. 240–301. Williams & Wilkins, Baltimore, Maryland, 1961.

58. Hall, P. F., *in* "The Androgens of the Testis" (K. B. Eik-Nes, ed.), pp. 73–115. Dekker, New York, 1970.

59. Hall, P. F., *in* "The Testis" (A. D. Johnson, W. R. Gomes, and N. L. Van Demark, eds.), Vol. 2, pp. 1–71. Academic Press, New York, 1970.

60. Harding, L. K., *Int. J. Radiat. Biol.* **3**, 539 (1961).

61. Hartree, E. F., and Srivastava, P. N., *J. Reprod. Fert.* **9**, 47 (1965).

62. Hollinger, M. A., *Life Sci.* **9**, 533 (1970).

63. Hotta, Y., and Stern, H., *Develop. Biol.* **26**, 87 (1971).

64. Hotta, Y., and Stern, H., *Nature (London), New Biol.* **234**, 83 (1971).

65. Howell, S. H., and Stern, H., *J. Mol. Biol.* **55**, 357 (1971).

66. Huckins, C., *Anat. Rec.* **169**, 533 (1971).

67. Huckins, C., *Cell Tissue Kinet.* **4**, 139 (1971).

68. Huckins, C., *Cell Tissue Kinet.* **4**, 313 (1971).

69. Ingles, C. J., and Dixon, G. H., *Proc. Nat. Acad. Sci. U.S.* **58**, 1011 (1967).

70. Jergil, B., and Dixon, G. H., *J. Biol. Chem.* **245**, 425 (1970).

71. Johnson, A. D., Gomes, W. R., and Van Demark, M. L., *in* "The Testis" (A. D. Johnson, W. R. Gomes, and M. L. Van Demark, eds.), Vols. 1–3. Academic Press, New York, 1970.

72. Jost, A., *in* "The Human Testis" (E. Rosenberg and C. A. Paulsen, eds.), pp. 11–18. Plenum, New York, 1970.

73. Kuehl, F. A., Patanelli, D. J., Tarnoff, J., and Humes, J. L., *Biol. Reprod.* **2**, 154 (1970).

74. Lam, D., and Bruce, W. R., *J. Cell Physiol.* **78**, 13 (1971).

75. Lam, D., Furrer, R., and Bruce, W. R., *Proc. Nat. Acad. Sci. U.S.* **65**, 192 (1970).

76. Leblond, C. P., and Clermont, Y., *Amer. J. Anat.* **90**, 167 (1952).

77. Lee, L. P. K., and Fritz, I. B., *J. Biol. Chem.* **247**, 7956. (1972).

77a. Lee, L. P. K., and Fritz, I. B., unpublished observations.

78. Liao, S., and Fang, S., *Vitam. Horm. (New York)* **27**, 17 (1969).

79. Lin, C. H., and Fritz, I. B., *Can. J. Biochem.* **50**, 963 (1972).

80. Ling, V., and Dixon, G. H., *J. Biol. Chem.* **245**, 3035 (1970).

81. Ling, V., Jergil, B., and Dixon, G. H., *J. Biol. Chem.* **246**, 1168 (1971).

82. Ling, V., Trevithick, J. R., and Dixon, G. H., *Can. J. Biochem.* **47**, 51 (1969).

83. Loir, M., and Wyrobek, A., *Exp. Cell Res.* **75**, 261 (1972).

84. Lostroh, A. J., *Acta Endocrinol. (Copenhagen)* **43**, 592 (1963).

84a. Lyon, M. F., personal communication.

85. Lyon, M. F., and Hawkes, S. G., *Nature (London)* **227**, 1217 (1970).

86. Males, J. L., and Turkington, R. W., *J. Biol. Chem.* **245**, 6329 (1970).

87. Males, J. L., and Turkington, R. W., *Endocrinology* **88**, 579 (1971).

172   IRVING B. FRITZ

88. Mancini, R. E., Alonso, A., Barquet, J., Alvarez, B., and Nemirovsky, M., *J. Fert. Reprod.* **8**, 325 (1964).
89. Mann, T., ed., "Biochemistry of Semen and of the Male Reproductive Tract." Wiley, New York, 1964.
90. Marquis, N. R., and Fritz, I. B., *J. Biol. Chem.* **240**, 2197 (1965).
91. Marushige, K., and Dixon, G. H., *Develop. Biol.* **19**, 397 (1969).
92. Marushige, K., Ling, V., and Dixon, G. H., *J. Biol. Chem.* **244**, 5953 (1969).
93. Means, A. R., in "The Human Testis" (E. Rosenberg and C. A. Paulsen, eds.), pp. 301–313. Plenum, New York, 1970.
94. Means, A. R., *Endocrinology* **89**, 981 (1970).
95. Means, A. R., and Hall, P. F., *Biochemistry* **8**, 4293 (1969).
96. Means, A. R., Hall, P. F., Nicol, L. W., Sawyer, W. H., and Baker, C. A., *Biochemistry* **8**, 1488 (1969).
97. Meistrich, M. L., *J. Cell. Physiol.* **80**, 299 (1972).
98. Meistrich, M. L., Bruce, W. R., and Clermont, Y., *Exp. Cell Res.* **79**, 213–227 (1973).
99. Meistrich, M. L., and Eng, V. W. S., *Exp. Cell Res.* **70**, 237 (1972).
100. Menon, K. M. J., and Smith, M., *Biochemistry* **10**, 1186 (1971).
101. Miller, R. G., and Phillips, R. A., *J. Cell. Physiol.* **73**, 191 (1969).
102. Moens, P., and Go, V. L. W., *Z. Zellforsch. Mikrosk. Anat.* **127**, 201 (1972).
103. Mohri, H., Mohri, T., and Ernster, L., *Exp. Cell Res.* **38**, 217 (1965).
104. Monesi, V., *J. Cell Biol.* **14**, 1 (1962).
105. Monesi, V., *Exp. Cell Res.* **39**, 197 (1965).
106. Monesi, V., *J. Reprod. Fert., Suppl.* **13**, 1 (1971).
107. Monn, E., Desautel, M., and Christiansen, R. O., *Endocrinology* **91**, 716 (1972).
108. Moses, M. J., *Annu. Rev. Genet.* **2**, 363 (1968).
109. Murad, F., Strauch, S., and Vaughan, M., *Biochim. Biophys. Acta* **177**, 591 (1969).
110. Murphy, H. D., *Proc. Soc. Exp. Biol. Med.* **118**, 1202 (1965).
111. Nicander, L., *Z. Zellforsch. Mikrosk. Anat.* **83**, 375 (1967).
112. Nishimune, Y., and Komatsu, T., *Exp. Cell Res.* **75**, 514 (1972).
113. Oakberg, E. F., *Amer. J. Anat.* **99**, 507 (1956).
114. Oakberg, E. F., *Anat. Rec.* **169**, 515 (1971).
115. Peters, H., *Phil. Trans. Roy. Soc. London, Ser. B* **259**, 91 (1970).
116. Peterson, R. N., and Freund, M., *Fertil. Steril.* **21**, 151 (1970).
117. Reissenweber, N. J., *Histochemie* **21**, 73 (1970).
118. Reissenweber, N. J., Arbes-Navarrette, I., and Sala, M. A., *Histochemie* **21**, 84 (1970).
119. Rivarola, M. A., Podesta, E. J., and Chemes, H. E., *Endocrinology* **91**, 537 (1972).
120. Rommerts, F. F. G., Cooke, B. A., van der Kemp, J. W. C. M., and van der Molen, H. J., *FEBS Lett.* **24**, 251 (1972).

121. Roosen-Runge, E. C., *Biol. Rev. Cambridge Phil. Soc.* **37**, 343 (1962).
122. Salisbury, G. W., and Lodge, G. R., *Advan. Enzymol.* **24**, 35 (1962).
123. Seiguer, A. C., and Castro, A. E., *Biol. Reprod.* **7**, 31 (1972).
124. Setchell, B. P., *J. Reprod. Fert.* **19**, 391 (1969).
125. Setchell, B. P., *in* "The Testis" (A. D. Johnson, W. R. Gomes, and N. L. Van Demark, eds.), Vol. 1, pp. 101–239. Academic Press, New York, 1970.
126. Setchell, B. P., *J. Reprod. Fert.* **23**, 79 (1970).
127. Setchell, B. P., Voglmayr, J. K., and Waites, G. M. H., *J. Physiol. (London)* **200**, 73 (1969).
128. Setchell, B. P., and Waites, G. M. H., *J. Reprod. Fert., Suppl.* **13**, 15 (1971).
129. Smith, P. E., *J. Amer. Med. Ass.* **88**, 158 (1927).
130. Smith, P. E., *Amer. J. Anat.* **45**, 205 (1930).
131. Srivastava, P. N., Adams, C. E., and Hartree, E. F., *J. Reprod. Fert.* **10**, 61 (1965).
132. Stambaugh, R., and Buckley, J., *Science* **161**, 585 (1968).
133. Steinberger, E., *Physiol. Rev.* **51**, 1 (1971).
134. Steinberger, E., and Nelson, W. O., *Endocrinology* **56**, 429 (1955).
135. Steinberger, E., Steinberger, A., and Ficher, M., *Recent Progr. Horm. Res.* **26**, 547 (1970).
136. Stern, H., and Hotta, Y., *Curr. Top. Develop. Biol.* **3**, 37 (1689).
137. Stern, H., and Hotta, Y., *in* "Handbook of Molecular Cytology" (A. Lima-de-Faria, ed.), pp. 521–531. North-Holland Publ., Amsterdam, 1970.
138. Sung, M. T., and Dixon, G. H., *Proc. Nat. Acad. Sci. U.S.* **67**, 1616 (1970).
139. Szego, C. M., *Advan. Cyclic Nucleotide Res.* **1**, 541 (1972).
140. Szego, C. M., Seeler, B. J., Steadman, R. A., Hill, D. F., Kimura, A. K., and Roberts, J. A., *Biochem. J.* **123**, 523 (1971).
141. Tepperman, J., *Perspect. Biol. Med.* **4**, 445 (1961).
142. Thompson, D. W., *in* "Aristotle's Works" (J. A. Smith and W. R. Ross, eds.), Vol. 4, p. 631. Oxford Univ. Press (Clarendon), London and New York, 1910.
143. Utakoji, T., *Methods Cell Physiol.* **4**, 1–17 (1970).
144. Utakoji, T., Muramatsu, M., and Sugano, H., *Exp. Cell Res.* **53**, 447 (1968).
145. Van Demark, M. L., and Free, M. J., *in* "The Testis" (A. D. Johnson, W. R. Gomes, and M. L. Van Demark, eds.), Vol. 3, pp. 233–312. Academic Press, New York, 1970.
146. Vernon, R. G., Go, V. L. W., and Fritz, I. B., *Can. J. Biochem.* **49**, 761 (1971).
147. Waites, G. M. H., *in* "The Testis" (A. D. Johnson, W. R. Gomes, and M. L. Van Demark, eds.), Vol. 1, pp. 241–279. Academic Press, New York, 1970.

148. Waites, G. M. H., and Setchell, B. P., *Advan. Reprod. Physiol.* **4**, 1 (1969).

149. Williams-Ashman, H. G., and Reddi, A. H., *in* "Biochemical Actions of Hormones" (G. Litwack, ed.), Vòl. 2, pp. 257–294. Academic Press, New York, 1972.

150. Wilson, J. D., *N. Engl. J. Med.* **287**, 1284 (1972).

151. Winer, A. D., and Nikitovitch-Winer, M. B., *FEBS Lett.* **16**, 21 (1971).

152. Yokoe, Y., and Hall, P. F., *Endocrinology* **86**, 18 (1970).

153. Zinkham, W. H., Blanco, A., and Clowry, C. J., *Ann. N.Y. Acad. Sci.* **121**, 571 (1964).

Enzyme Degradation and Its Regulation by Group-Specific Proteases in Various Organs of Rats

NOBUHIKO KATUNUMA

Department of Enzyme Chemistry
Institute for Enzyme Research
School of Medicine
Tokushima University
Tokushima, Japan

I. Introduction

Almost thirty years ago studies with ^{131}I-labeled serum albumin demonstrated that serum proteins are degraded *in vivo*. Since then, with the aid of more sophisticated isotope techniques and the development of immunological techniques for the selective precipitation of specific proteins from crude tissue extracts, it was established that most tissue proteins undergo continual synthesis and breakdown, albeit at different rates.

Schimke (*16*) emphasized that the steady-state level of a given enzyme is determined by the relative rates of its synthesis and degradation and demonstrated that these rates were independently controlled by dietary and hormonal factors. It is obvious,

175

therefore, that rigorous control of both biosynthesis and degradation of a particular enzyme is essential to maintain it at a fixed intracellular concentration. Whereas considerable progress has been made in understanding the mechanism of enzyme synthesis and its regulation at the gene level, the mechanism of enzyme degradation and its control are only poorly understood (*9, 11, 12, 17, 18*). De Duve (*3*) proposed that lysosomal proteases (also called cathepsins) may play an important role in the degradation of enzyme protein. This is an attractive possibility in view of the fact that cathepsins are released into the cytoplasm upon disruption of the lysosomal membrane and, furthermore, that protein can be taken into the lysosome under certain physiological conditions (*2*). Nevertheless, the view that all intracellular enzymes are degraded by the lysosomal proteases is difficult to reconcile with the following considerations: (1) The pH optimum of lysosomal proteases lies between 4.5 and 5.5; therefore they are not well designed for the degradation of most enzymes at physiological pH. (2) Lysosomal proteases exhibit little substrate specificity, therefore they lack the capacity to catalyze the selective degradation of a particular enzyme in the presence of others. The activity of lysosomal proteases does not appear to be altered by variations in dietary or hormonal states known to affect the rates of specific enzyme synthesis and degradation. For these reasons, it seems that lysosomal proteases lack the characteristics that are needed to account for the rapid, complete, *in vivo* degradation of a specific enzyme or groups of enzymes, in response to variations in the metabolic state of the organism.

Schimke *et al.* (*1*) have recently obtained evidence that protein degradation *in vivo* may occur in a stepwise manner. By means of a successive double-labeling technique, they showed that the first step involves limited breakdown of the protease into a few relatively large fragments, which are subsequently degraded further into low molecular weight derivatives. It was proposed that cells contain a special protease that initiates degradation of native enzymes by catalyzing very limited proteolysis, producing relatively large molecular fragments that are ultimately degraded completely by other proteases. Whereas this proposal is attractive in principle, it seems unlikely that a single protease catalyzes the first step in this breakdown of all native enzyme. The protease that initiates protein breakdown must have a high degree of specificity to account for the differential breakdown of particular enzymes *in vivo* and the regulation of this breakdown by nutritional factors and hormones.

With regard to this consideration, several proteases that exhibit the desired substrate specificity and response to growth conditions were recently discovered in the rat. One class of protease catalyzes limited proteolysis of only the apo form of pyridoxal phosphate-dependent enzymes; however, it cannot attack the apo form of NAD-dependent enzymes or other non-coenzyme-dependent enzymes tested. This new protease is referred to as the *group-specific protease for pyridoxal enzymes (6, 13)*. Since these group-specific proteases catalyze only limited proteolysis of their substrate enzymes, it is proposed that their regulatory function is concerned with the inactivation of most, but not all, pyridoxal enzymes and the initiation of their degradation. In addition to the proteases specific for pyridoxal enzymes, two other classes of group-specific proteases have been discovered in animal tissues. One of these catalyzes partial hydrolysis of the apoenzyme forms of FAD-dependent dehydrogenases, whereas the other is specific for the apoenzyme forms of various NAD-dependent dehydrogenases (8). Holzer and Katsunuma (5) have also demonstrated the existence of similar group-specific proteases in yeast. At present, it is believed that each class of group-specific proteases is concerned with initiation of the degradation of their respective apoenzyme substrates. After the primary attack by the group-specific protease, it is presumed that ultimate degradation of pyridoxal enzymes to amino acids is achieved by lysosomal or other nonspecific proteases. It is reasonable to suggest that there is a separate group-specific protease for each class of apoenzymes. In this case, the concentrations of all apoenzymes in a given class can be regulated by only one protease, or at most a few. This avoids the uneconomical necessity of having a specific protease to initiate degradation of each individual enzyme. Presumably, within each class, regulation of the degradation rate (and hence the intracellular concentration) of various enzymes could be achieved by differences in their affinities for the common coenzyme as well as their substrates or other effectors.

It is the purpose of this presentation to summarize knowledge concerning the systems involved in the intracellular degradation of selected enzyme proteins. In particular the role of group-specific proteases in regulating the intracellular levels of vitamin B_6-dependent enzymes is discussed.

II. Biological Background of the Discovery of Group-Specific Protease for Pyridoxal Enzymes

Discovery of the pyridoxal enzyme-specific protease was an outgrowth of investigations on the effects of vitamin B_6-deficient and nonprotein diets on the tissue levels of pyridoxal enzymes. It was found that the levels of ornithine transaminase (OTA) and aspartate transaminase (GOT) in the small intestine of rats decreased markedly when the animals were maintained on vitamin B_6-deficient diet for 4 weeks. As illustrated in Fig. 1, the decrease in these enzyme levels could be reversed within 4–5 hours by injection of a low dose of vitamin B_6 derivatives. In contrast, the levels of non-B_6-dependent enzymes, i.e., leucine aminopeptidase (LAP) and glutaminase, were unaffected by these treatments. The differential effect of pyridoxine derivatives on the tissue levels of pyridoxal enzymes suggested that the coenzyme either stimulated the biosynthesis of the apoenzymes or protected them from the degradation.

To examine these possibilities, the effect of B_6 administration on the differential synthesis of these enzymes compared to total cellular protein was determined. The *in vivo* incorporation of leucine-^{3}H into protein was used as a measure of protein synthesis. In order to determine the amount of leucine-^{3}H incorporated into the specific enzymes, they were selectively separated from other cellular proteins by means of immunological techniques. Data summarized in Table I show that after B_6 administration the amount of leucine-^{3}H incorporated into ornithine transaminase is no greater than the average amount incorporated into the total protein fraction. Therefore B_6 does not increase the rate of synthesis of ornithine transaminase. The results suggest, instead, that the increase in ornithine transaminase activity caused by B_6 administration could be due to the ability of coenzyme (pyridoxal phosphate) to protect this enzyme from degradation (7).

This conclusion was verified by the demonstration that cell-free

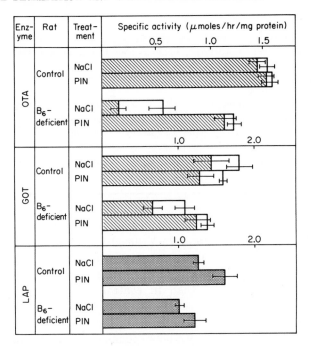

Fig. 1. Effect of pyridoxine injection on the activity of several enzymes of small intestine of vitamin B₆-deficient rats. Rats weighing 50 gm were maintained on a pyridoxine-deficient diet for 40 days before the experiments. Bilateral adrenalectomy was performed at 48 hours before the experiments. Animals were injected intraperitoneally with either 100 μg per 100 gm body weight of pyridoxine (PIN) or 0.9% saline solution 5 hours before sacrifice. Homogenates of various organs (20% w/v) were prepared in 0.05 M potassium phosphate buffer, pH 7.5, with a Potter-Elvehjem type of homogenizer. The homogenates were sonicated at 10 kc for 2 minutes and then centrifuged for 10 minutes at 10,000 g. The supernatant was used as the enzyme source.

extracts of the small intestine contained a specific protease that catalyzes partial hydrolysis of apo-ornithine transaminase and the apo form of other pyridoxal enzymes; moreover, degradation of these enzymes by the new protease is prevented by the presence of pyridoxine derivatives.

Khairallah and Pitot (*10*) had reached similar conclusions with regard to the *in vivo* degradation of serine dehydratase. In these studies, serine dehydratase was prelabeled *in vivo,* and the effect

TABLE I
IMMUNOCHEMICAL ANALYSIS OF THE INCREASE IN ORNITHINE TRANS-
AMINASE (OTA) ACTIVITY, INCREASE BY PRYIDOXINE (PIN) IN
SMALL INTESTINE OF B$_6$-DEFICIENT RATS

Treatment	OTA specific activity (440 mU/hr/ mg protein)	Soluble protein (cpm/gm tissue)	OTA (cpm/gm tissue)	OTA soluble protein $\times 10^{-2}$
NaCl	0.59 ± 0.03	$15,025 \pm 651$	681 ± 31	4.5 ± 0.02
PIN	0.85 ± 0.04	$14,247 \pm 2336$	617 ± 38	4.6 ± 1.30

of pyridoxine administration on degradation of the labeled dehydratase was then determined. Although all animals exhibited a
3-fold increase in total enzyme activity over a 9-hour period, the
decrease in specific radioactivity was much slower in animals receiving pyridoxine injections than in control animals injected with
saline solution only. From semilogarithmic plots of the data (Fig.
2) it was calculated that the half-life of serine dehydratase in pyri-
doxine-treated animals was 9.3 hours compared to 4.2 hours for animals in the control group.

The regulation of intracellular apoenzyme levels by coenzymes
as a stabilization effect was also discussed by Greengard (4).

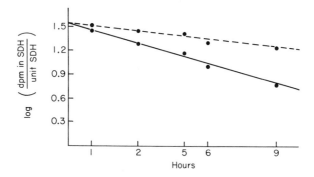

FIG. 2. Effect of pyridoxine on the half-life of serine dehydrase. ●---●,
Pyridoxine, $t_{1/2} = 9.3$; ●——● Nacl, $t_{1/2} = 4.3$. From Khairallah and Pitot
(10).

III. Organ Distribution and Multiplicity of Group-Specific Protease for Pyridoxal Enzymes

The small intestine contains two pyridoxal enzyme-specific proteases; one in localized in the muscle layer and the other in the mucosa layer. Similar proteases are present also in rat liver and in skeletal muscle (*1*).

All these proteases have a relative substrate specificity for pyridoxal-dependent enzymes, and all have an alkaline pH optimum; however, the proteases from different organs respond differently to variations in nutritional and hormonal conditions, and they differ in their specific activities in the presence of various substrates, and in their physical-chemical characteristics (viz, molecular weights and elution profiles during anion exchange chromatography). Therefore, these proteases might be regarded as isoenzymes that are functionally concerned with the regulation of pyridoxal-dependent enzyme concentrations. The observed differences in specific activities and responses to hormonal and nutritional effectors might reflect differences in the metabolic functions of the organs in which they are situated. For example, the great ability of the skeletal muscle protease to degrade muscle phosphorylase, and the fact that its activity is increased by nonprotein diets or by starvation or by B_6 deficiency, or by amputation of a control nerve, suggests that it may have an important role in the regulation of muscle metabolism. Comparison of these group-specific proteases from different organs is described in the following subsections.

A. Group-Specific Protease in Skeletal Muscle

1. PURIFICATION AND PROPERTIES

Skeletal muscle prepared from rats fed high-protein diets or nonprotein diets was used as a starting material for purification of the pyridoxal enzyme protease. A method which achieves 9682-fold purification of the protease is summarized in Table II. The work on group specific protease in skeletal muscle has been carried out by K. Suzuki, K. Chichibu, T. Katsunuma, T. Shiotani, E.-G Afting, and N. Katunuma (*6, 19*). The large increase in total activity ob-

TABLE II

PURIFICATION OF ORNITHINE TRANSAMINASE INACTIVATING
ENZYME FROM MUSCLE

Procedure	Volume (ml)	Total activity (units)	Total protein (mg)	Specific activity (units/mg)	Purity	Recovery (%)
Crude extract	5000	57,700	105,900	0.55	1.0	100
Acetone powder extract (with 0.05 M, pH 8.0 KPB)	2500	67,000	32,800	2.04	3.7	116
Salmine sulfate treatment	2500	107,100	30,800	3.48	6.4	186
Acetone 0–40% precipitate	1200	72,000	7,800	10.18	18.7	125
Hydroxyapatite column	10.3	61,000	301	202.9	372.3	106
Sephadex G-75 column	5.5	22,100	4.15	5325.3	9682.4	38

served in step 4 (acetone precipitation step) of this procedure is probably due to the removal of endogenous inhibitory material. In fact, a heat-stable protein that inhibits the protease has been partially purified from skeletal muscle.

As judged by a column chromatography on Sephadex G-75 (Fig. 3) and by acrylamide gel disc electrophoresis, the most highly purified enzyme preparation is a homogeneous protein and has a molecular weight of approximately 12,000–14,000. The protease is a basic protein, since it migrates toward the cathode during electrophoresis at pH 6.9 and is not absorbed on DEAE-cellulose ion exchange.

Data in Fig. 4 show that inactivation of ornithine transaminase by the purified protease is a first-order reaction with respect to the transaminase concentration; this is because the substrate is not saturating under standard assay conditions. One unit of protease is therefore defined as the amount required to catalyze 50% inactivation of the substrate (ornithine transaminase) under standard assay conditions. The pH optimum of the protease is dependent upon the substrate used. As shown in Fig. 5, the optimum for inactivation of ornithine transaminse is at pH 9.0, whereas pH 8.0 is the

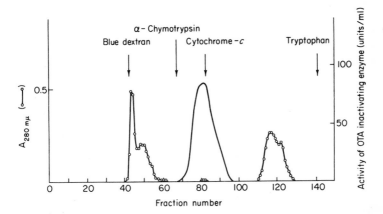

FIG. 3. Column chromatography of group-specific protease from muscle on Sephadex G-75. The enzyme was eluted from the column with 0.05 M potassium phosphate buffer, pH 8.5, at flow rate of 20 ml/hour. Fractions of 1 ml were collected and assayed for the absorbance at 280 mμ and for group-specific protease activity. The heavy line indicates group-specific protease activity. Arrows indicate the positions at which reference compounds appeared.

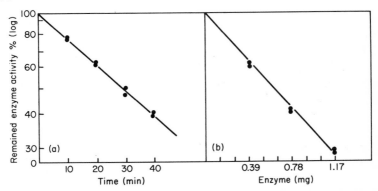

FIG. 4. Time course (a) and dose response curve (b) for group-specific protease in muscle. The reaction mixture (final volume, 0.3 ml) contained 70 μmoles of potassium phosphate buffer, pH 8.5, 40–60 units (0.1 mg) of substrate enzyme, and a suitable amount of the group-specific proteases preparation. The enzyme was incubated at 37°C, and the reaction was stopped by 10-fold dilution with cold buffer. Then the remaining activity of the substrate enzyme was assayed, and the percentage of inactivation was calculated. One unit of the enzyme is defined as the amount inactivating 50% of the substrate enzyme in 30 minutes under these conditions.

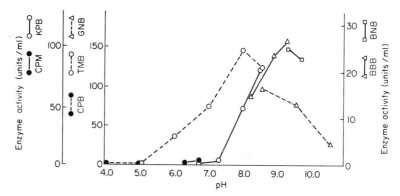

FIG. 5. Optimum pH of group-specific protease from muscle. Solid line indicates optimum pH curve with ornithine transaminase as a substrate, and the following buffers were used: ●——●, citrate phosphate (CPB); ○——○, potassium phosphate (KPB); △——△, is glycine-NaOH (GNB). □——□, borax-NaOH (BNB). Dotted line illustrates optimum pH curve with *N*-acetyl-L-tyrosine ethyl ester as a substrate, and the following buffers were used: ●---●, citrate phosphate (CPB); ○---○, Tris-maleate (TMB); △---△, glycine NaOH (GNB).

optimum for hydrolysis of the synthetic substrate *N*-acetyl-L-tyrosine ethyl ester (ATEE).

In contrast to lysosomal protease, the enzyme from skeletal muscle is inhibited by diiosopropyl fluorophosphate (DFP) and is not affected by sulfhydryl reagents; therefore it belongs in the class of seryl-OH proteolytic enzymes.

Among various metal ions tested, in order of their effectiveness, Ca^{2+}, Zn^{2+}, and Mn^{2+} inhibited the protease, whereas Mg^{2+}, K^+, and Na^+ were without effect. Addition of $3 \times 10^{-4}\ M\ Ca^{2+}$ caused 50% inhibition.

2. SUBSTRATE SPECIFICITY

As shown in Table III, the skeletal muscle protease exhibits relative specificity for the apo forms of pyridoxal-dependent enzymes as substrates. Ornithine transaminase, homoserine deaminase, serine dehydratase, and muscle phosphorylase are all rapidly inactivated. In contrast, nonpyridoxal-dependent enzymes are not inactivated, even upon prolonged incubation with the protease. Moreover, a good synthetic substrate for trypsin, *p*-toluene sul-

TABLE III
Substrate Specificity (Enzyme Substrates)

Substrate	Source	Concentration (units/tube)	Percent activity (%)
Ornithine transaminase	Rat liver	0.6	100
Phosphorylase[a]	Rabbit muscle	0.6	108
Homoserine deaminase	Rat liver	0.7	74
Serine dehydratase	Rat liver	0.8	56
Aspartate transaminase	Rat liver	0.6	0
Malic dehydrogenase	Rat kidney	8.8	0
Glutamic dehydrogenase	Bovine liver	17.2	0
Lactic dehydrogenase	Rabbit (4H)	14.5	0
Glyceroaldehyde dehydrogenase	Rat liver	10.0	0
Phosphate-independent glutaminase	Rat kidney	1.0	0
Urease	Jack bean	0.22	0

[a] Holophosphorylase.

fonyl-L-arginine methyl ester (TAME) is not attacked by the protease, and the synthetic substrate for chymotrypsin, N-acetyl-L-tyrosine ethyl ester (ATEE) is hydrolyzed only 30% as rapidly by the protease as is ornithine transaminase. L-tyrosine ethyl ester (TEE) is not attacked by the protease.

Although ATEE is a fairly good substrate for purified muscle protease, this is not sufficient reason to classify the protease as a chymotrypsin-type enzyme, especially in view of the fact that high concentrations of the chymotrypsin inhibitor, p-aminophenyl-β-phenyl propionate, do not affect the activity of the muscle protease when either ATEE or ornithine transaminase are used as substrate. Except for muscle phosphorylase, which is inactivated even in the holo form, although the apo form of the phosphorylase is affected ten times more rapidly than the holo form, only the apo forms of the pyridoxal-dependent enzymes are attacked by the group-specific protease. Either pyridoxal phosphate or pyridoxamine phosphate can protect all the enzymes except phosphorylase from inactivation, whereas pyridoxine phosphate or pyridoxamine are without effect. Since pyridoxal phosphate does not inhibit the protease-catalyzed hydrolysis of phosphorylase or the synthetic substrate ATEE, it may be concluded that inhibition by pyridoxal phosphate is caused by its reaction with the apoenzyme substrate, rather than with the protease itself. This suggests that the pyridoxal-dependent enzymes are vulnerable to attack by the protease only when the coenzyme binding site is vacant. It is noteworthy that muscle contains a heat-stable protein factor which inhibits the group specific protease in that tissue. This factor has been partially purified from rat muscle homogenates by means of heat treatment (80°C for 1 minute) followed by chromatography on Sephadex G-25. This factor does not inhibit the activity of trypsin when TAME is the substrate. The possible role of this inhibitory protein in the regulation of the group-specific protease has not been investigated.

3. Control of the Group-Specific Protease by Physiological Conditions

The concentration of the pyridoxal enzyme-specific protease varies greatly from one muscle to another. As shown in Fig. 6, the activity in red muscle is much higher than in white muscle, and no activity could be detected in the heart muscle. Levels of the protease varied also as a function of the dietary

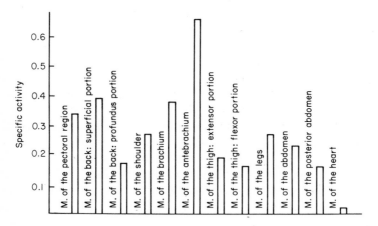

Fig. 6. Localization of group-specific protease activity in muscles. The animals, male Wister strain rats, were maintained on laboratory chow. Specific activities of the muscle protease extracted in 0.2 M potassium phosphate buffer pH 8.5 were measured according to standard assay system.

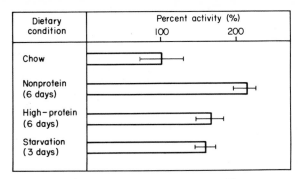

Fɪɢ. 7. Effect of dietary conditions on group-specific protease activity in rat skeletal muscle. Wistar strain rats, weighing about 200 g, were maintained on nonprotein diet for 6 days and on starvation for 3 days before experiments. Specific activities in the homogenate of skeletal muscle in 0.2 M potassium phosphate buffer, pH 8.5, were measured with the standard assay system as described under Experimental Procedure. Relative activities were compared to 100% specific activity (mean 0.18 unit/mg) of the protease in muscles of rats fed on laboratory chow.

condition. Data in Fig. 7 show that the level in muscle from rats fed on ordinary laboratory chow is considerably lower than that in muscles from animals fed on either nonprotein or high-protein diets, or in animals maintained on a starvation diet. The enhancement of group-specific protease activity in response to starvation or nonprotein diet is consistent with the well-known fact that under these conditions there is an increase in the amount of amino acids released and transported from muscle into the blood.

A physiological role of the group-specific protease in the breakdown of muscle protein is further indicated by the observation that there is a significant increase in the level of this enzyme during atrophy of the gastrocnemius muscle caused by amputation of the ischiadicus nervus. This was established by an experiment in which only one side of the nervus ischiadicus was amputated at the position of the foramen infrapiriforme. Fifteen days after the operation, the muscle weight of the gastrocnemius muscle on the amputated side was only 50% that found on the normal side, but the level of group-specific protease on the amputated side was 3 times higher than that on the normal side. This result agrees with the data show-

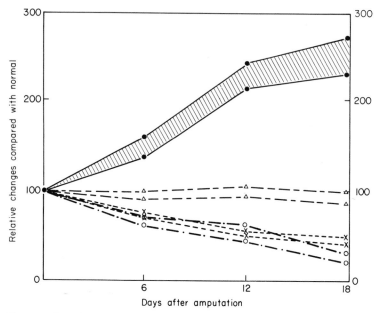

FIG. 8. Changes of group-specific protease and muscle phosphorylase in gastrocnemius and soleus by amputation of the nervus ischiadicus. The following symbols are used: ●——●, group-specific protease; ○-•-•-○, muscle phosphorylase; ×---×, muscle weight; △-•-•-△, aspartate transaminase. Relative changes were compared with 100% of normal side.

ing that amputation of the nervus ischiadicus leads to a marked decrease in the level of muscle phosphorylase, as shown in Fig. 8.

Figure 9 shows the changes in the protease activity, muscle weight, phosphorylase activity, ornithine transaminase activity in gastrocnemius and soleus on the cause of vitamin B_6 deficiency.

B. Group-Specific Protease in Small Intestine

1. PREPARATION AND PURIFICATION (6)

The intestinal contents include large amounts of exocrine juice which is secreted from the pancreas, intestine, and bile duct; therefore a thorough washing of the starting material is an essential step in the preparation of enzymes from the small intestine. The follow-

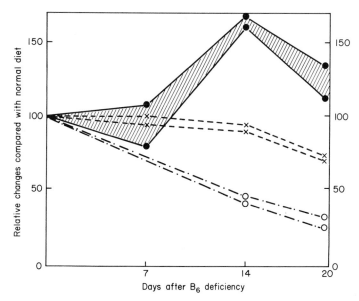

Fig. 9. Changes of group-specific protease and muscle phosphorylase in gastrocnemius and soleus during vitamin B₆ deficient diet intake. The following symbols are used: ●——●, group-specific protease; ○-·-·-·○, muscle phosphorylase; ✕ - - - ✕, muscle weight. Relative changes were compared with 100% of normal diet rats.

ing washing procedure provides suitable starting material for the isolation of group-specific proteases from the mucosa and muscle layers of the small intestine. The intestinal rumen is first extensively washed with 0.15 M KCl solution until all contents are removed. The intestines are dissected and rinsed in a beaker several times with 0.15 M KCl. When activity can no longer be detected in the washing fluid, the mucosa and muscle layers are separated by scratching with a spatula. Homogenates of each tissue serve as starting material for the isolation of the group-specific proteases. Studies on the proteases in small intestine have been carried out by Kominami, Kobayashi, Hamaguchi, and Katunuma (6).

The procedures used in the purification of the group-specific protease from the muscle and mucosa layers are shown in Tables IV and V, respectively.

TABLE IV

PURIFICATION OF GROUP-SPECIFIC PROTEASE FROM SMALL
INTESTINE (MUSCLE)

Procedure	Total activity (units)	Specific activity (units/mg)	Purity
Extract (0.15 M KCl)	21,250	2.4	1.0
(NH₄)₂SO₄ precipitation (0–75%)	34,000	7.7	3.2
Acetone (30–55%, v/v)	29,460	18.7	7.8
DE-52 (0.005 M, pH 8.0 eluents)	8,824	37.4	15.6
CM-Sephadex (0.05 M, pH 7.0 KPB)	7,070	70.0	29.2
Sephadex G-75 (0.05 M, pH 7.5 KPB)	6,435	71.5	29.8
Crystallization by (NH₄)₂SO₄	—	75.0	31.3

After preliminary fractionation of the crude muscle layer extract
by steps involving precipitation with ammonium sulfate and acetone, the partially purified enzyme was absorbed on a DEAE-cellulose (DE-52, Whatman) column equilibrated with 0.005 M potassium phosphate buffer (pH 8.0) and was eluted stepwise with 0.25
M to 0.8 M KCl solution. The specific protease from the muscle
layer of the small intestine eluted in 0.005 M buffer. Further purification was obtained by column chromatography on Sephadex
G-75. The elution profile of the muscle layer enzyme is illustrated
in Fig. 10. An identical peak of activity was obtained whether
ornithine transaminase or the synthetic substrate (ATEE) was used
as the assay substrate; this suggested that a single protease attacks
both substrates. This procedure led to a 31-fold enrichment of protease activity from the crude extract. The enzyme was crystallized
by ammonium sulfate (Fig. 11).

A similar procedure was used in purification of the group-specific
protease from extracts of the mucosa layer of the small intestine.
The results are shown in Table V. In contrast to the behavior of
protease from the muscle layer, the enzyme from the mucosa layer
is eluted from DEAE-cellulose (DE-52, Whatman) columns by
a relatively high concentration of salt (0.25 M KCl). The procedure
for purifying the mucosa layer enzyme is summarized in Table
V. The procedure led to a 1200-fold purification of the enzyme.

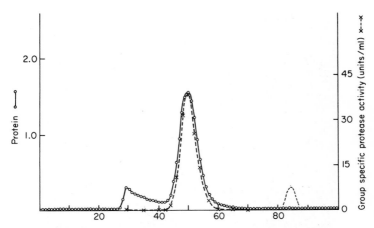

Fig. 10. Column chromatography of Sephadex G-75 of specific protease from the muscle layer of the small intestine. Elution conditions were as described in Fig. 3. The solid line expresses protease activity, and the dashed line indicates absorbance at 280 nm. . . . , (NH₄)₂SO₄.

TABLE V

PURIFICATION OF GROUP-SPECIFIC PROTEASE FROM
MUCOSA LAYER OF SMALL INTESTINE

Fraction	Volume (ml)	Total activity (units)	Total protein (mg)	Specific activity (units/mg)	Purity
Particulate fractions (Triton treatment, 4%)	1200	6540	14,400	0.45	1.0
1st acetone fractionation (0–75%)	310	4650	4,030	1.15	2.6
2nd acetone fractionation (30 ∼ 75%)	40	4000	520	7.7	17.1
DE-52 Column (eluted with 0.25 M KPB)	90	5400	99	64.0	142.2
Calcium phosphate gel (eluted with 0.2 M, pH 8.0 KPB)	17	5370	30.6	174.5	387.8
Sephadex G-75	47	2970	5.6	545	1211.1

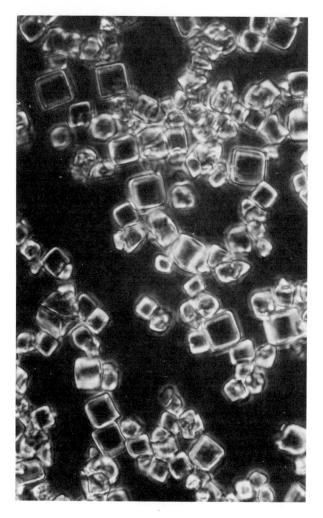

FIG. 11. Crystals of the group-specific protease from the muscle layer of the small intestine. ×400.

2. Some Properties of the Purified Intestinal Proteases

a. The Muscle Layer Protease. The crystallized enzyme prepared from the muscle layer was homogeneous as judged by electrophoresis on either cellulose membranes or acrylamide disc gel (Fig. 12).

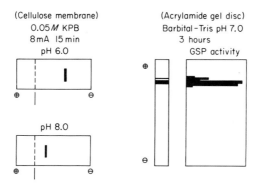

FIG. 12. Electrophoretic pattern of the protease from the muscle layer of the small intestine.

The molecular weight was approximately 23,000 as determined by gel filtration on Sephadex G-75. In contrast to most other tissue proteases that have been studied, both pyridoxal enzyme-specific proteases from the small intestine exhibit a pH optimum in the alkaline range (pH 8.6). Data in Table VI show that the muscle layer enzyme exhibits relative substrate specificity for the apoenzyme forms of pyridoxal-dependent enzymes, but the holo forms are resistant. Nonpyridoxal-dependent enzymes were not attacked, even with prolonged incubation or with 10 times as much protease as is needed to inactivate pyridoxal-dependent enzymes. Furthermore, p-toluene sulfonyl-L-arginine methyl ester (TAME), which is a good substrate for trypsin, is not hydrolyzed by the purified muscle layer enzyme. However, the enzyme readily hydrolyzes the synthetic substrates L-tyrosine ethyl ester (TEE) and N-acetyl-L-tyrosine ethyl ester (ATEE), both of which are good substrates for chymotrypsin. TEE was hydrolyzed at about the same rate as was ornithine transaminase, but ATEE was hydrolyzed about 10 times more rapidly than ornithine transaminase. Nevertheless, it was found with the analogous enzyme from skeletal muscle that the enzyme from the muscle layer of the small intestine was not inhibited by the chymotrypsin inhibitor, p-aminophenyl-β-phenyl propionate. The intestinal muscle enzyme is similar to the skeletal muscle enzyme in other respects also. It attacks only the apo forms of pyridoxal enzymes (except for the low activity with phosphorylase); moreover, pyridoxal phosphate and pyridoxamine phosphate inhibit

TABLE VI
SUBSTRATE SPECIFICITY OF MUSCLE LAYER
ENZYME IN SMALL INTESTINE

Enzyme	Activity[a] (%)
Ornithine transaminase	100
Serine dehydratase	19
Homoserine deaminase	59
Phosphorylase	22
Tyrosine transaminase	0
Holo-ornithine transaminase	4
Malic dehydrogenase	0
Glutamic dehydrogenase	0
Lactic dehydrogenase	0
Glyceroaldehyde dehydrogenase	0
Glutaminase (PI)	0
ATEE[b]	1368
TEE[c]	118
TAME[d]	0
Acetylcholine	0
Leucine-amide	0

[a] Values are the activities of each substrate enzyme expressed as percentages of that of ornithine transaminase.
[b] N-Acetyl-L-tyrosine ethyl ester.
[c] L-Tyrosine ethyl ester.
[d] p-Toluene sulfonyl-L-arginine methyl ester.

its ability to degrade ornithine transaminase but not its ability to degrade ATEE. However, in contrast to the high sensitivity of skeletal muscle protease to inhibition by Ca^{2+}, the intestinal enzyme was unaffected by even high concentrations (10^{-3} M) of Ca^{2+}. Finally, the catalytic turnover number of the muscle layer enzyme from the small intestine is only 1/50 that observed for the skeletal muscle enzyme.

From the data presented in Table VII, it is evident that the level of the muscle layer enzyme is significantly influenced by the diet. Growth on a nonprotein diet led to a substantial increase in the level of the enzyme compared to that obtained by growth on ordinary laboratory chow or growth on either a high protein diet or under fasting conditions.

TABLE VII
CHANGES OF GROUP-SPECIFIC PROTEASE ACTIVITY IN
MUSCLE LAYER OF SMALL INTESTINE UNDER VARIOUS
DIETARY CONDITIONS

Diet	Total activity (units)	Specific activity (units/mg)
Laboratory chow	31.57	0.29
Fasting (1 week)	16.60	0.44
High protein (1 week)	36.73	0.47
Nonprotein (1 week)	40.00	0.83

TABLE VIII
DISTRIBUTION OF GROUP-SPECIFIC PROTEASE IN
VARIOUS ORGANS

Site	Protease (units/mg)
Skeletal muscle	
M. antebrachium	0.7
M. of the back (profundus portion)	0.2
M. of the legs	0.3
Digestive tract	
Muscle layer of: Stomach	0.02
Duodenum	0.7
Jejunum	0.6
Ileum	0.6
Colon	0.1
Mucous layer of duodenum	0.05–0.1
Vena cava	0.2
Urinary bladder	$\doteqdot 0$
Uterus	0.02
Heart	$\doteqdot 0$
Liver (mitochondrial sonicate)	0.01–0.03

As shown in Table VIII, the concentration of the muscle layer enzyme varies greatly in each portion of the digestive tract.

b. *The Mucosa Layer Protease.* The molecular weight of the purified group-specific protease from the mucosa layer of the small intestine is about 25,000–30,000 as determined by gel filtration on Sephadex G-75. The characteristics of this enzyme are very similar to those isolated from skeletal muscle or from the muscle layer of the small intestine. All three enzymes have an alkaline pH optimum (ca pH 8.6). They all exhibit relative substrate specificity for the apo forms of pyridoxal-dependent enzymes, and their degradation of these subtrates is prevented by pyridoxal phosphate. Neither protease can hydrolyze the trypsin substrate TAME, but all three proteases hydrolyze ATEE, the synthetic substrate for chymotrypsin. They differ, however, in their abilities to hydrolyze ATEE as well as TEE, the other synthetic substrate for chymotrypsin. In comparison with their ability to hydrolyze the apoenzyme form of ornithine transaminase, ATEE is hydrolyzed ten times as fast by the muscle layer enzyme, 50% as fast by the mucosa layer enzyme, and 30% as fast by the skeletal muscle enzyme. However, only the mucosa layer enzyme is able to hydrolyze TEE.

C. Group-Specific Protease in Liver

Liver appears to play an important role in the turnover of pyridoxal enzymes. The work on the liver protease has been carried out by Y. Banno and N. Katunuma. A proteolytic activity capable of inactivating ornithine transaminase has been detected and partially purified from rat liver heavy mitochondrial fraction. The purification method is illustrated in Table IX.

The liver enzyme appears to differ from the group-specific proteases from other tissues in their behavior on column chromatography. In DEAE-cellulose (DE-52, Whatman) column chromatography, onithine transaminase inactivating activity is eluted with 0.75 M buffer. After sufficient protamine treatment, the supernatant was applied on Sephadex G-75. The liver enzyme appears just before

TABLE IX
PURIFICATION OF GROUP-SPECIFIC PROTEASE FROM LIVER

Procedure	Specific activity (units/mg)	Purification (fold)
Mitochondrial sonicate	0.03	1
Sonicate precipitate	1.03	34
Buffer extracts, 0.5 M	3.6	120
Concentration by 60% acetone	6.2	207
DEAE-cellulose column	89.5	2983
Protamine treatment, 50%	117.9	3930
Sephadex G-75	251.0	8367

cytochrome c on the column, so the molecular weight of the enzyme was assumed to be about 14,000. Since the liver enzyme binds strongly with nucleic acid, the behavior on column chromatography differs before and after complete separation from nucleic acid. The procedure illustrated in Table IX led to more than 8000-fold purification of the enzyme. The liver enzyme showed weak activity with ATEE as a substrate, but none of the enzyme showed activity with TAME, and the liver enzyme showed optimum activity at about pH 8.5. The purified liver enzyme also exhibits relative substrate specificity for the apo forms of pyridoxal enzymes. Ornithine transaminase, homoserine deaminase, and muscle phosphorylase, serine dehydratase are all inactivated in the same degree by the liver protease, but malic dehydrogenase, lactic dehydrogenase, tyrosine transaminase and glutamic dehydrogenase are not inactivated. Feeding a nonprotein diet causes 3- to 4-fold increases in liver enzyme activity, and regenerating liver and neonetal liver showed very low activities.

IV. Comparison of Group-Specific Proteases for Pyridoxal Enzymes from Various Organs

There have been few previous studies of intracellular proteases which act in the alkaline pH range and no previous reports on proteases which show specificity for special groups of enzyme proteins. We have obtained evidence for several such intracellular proteases, which share the following properties: (1) They normally exist in latent form and are converted to the active form by as yet undefined mechanisms. Thus the total activity may increase remarkably during purification. This may be related to the presence of specific inhibitors, protein in nature, in the tissues from which the proteases are isolated. (2) They all show relative specificity for pyridoxal enzyme proteins. (3) They show optimum activity between pH 8.5 and 9.0. (4) The apo forms of the pyridoxal enzymes are susceptible, but the holo forms are resistant. (5) In general, they show activity with synthetic chymotrypsin substrates, but not with synthetic trypsin substrates. (6) They are inhibited by DFP, but are unaffected by sulfhydryl reagents; they therefore can be classified as seryl, rather than as sulfhydryl, proteases. (7) Their activity responds to changes in various dietary conditions, but these responses may differ from organ to organ, as demonstrated for the muscle and intestinal proteases.

They may differ with respect to the following properties. (1) molecular weight, (2) elution patterns on DEAE column chromatography, (3) mobility on electrophoresis, (4) relative activity toward various aopyridoxal enzymes, (5) activity toward ATEE and TEE, (6) specific activity. On the basis of these similarities and differences, the group-specific proteases for pyridoxal enzymes may be regarded as a group of isoenzymes. The distribution of group-specific protease for pyridoxal enzymes in various organs are compared in Table VIII.

The similarities and differences in properties are summarized in Table X.

V. Other Studies on Intracellular Proteases That Affect Enzyme Proteins

Almost no work has been published on intracellular proteases that affect enzyme proteins in mammalian systems, except for a few studies on degradation of certain enzymes by lysosomal proteases. The inactivation of tryptophan synthetase in extracts of yeast grown under specific conditions was reported by Manney (15), and Katsunuma et al. (5) have purified two different enzymes that inactivate tryptophan synthetase. Among various yeast enzymes studied as substrates, the inactivating enzymes showed a relative specificity for pyridoxal enzymes, including threonine dehydratase, ornithine transaminase, and as aspartic acid transaminase; other types of enzyme tested were not inactivated. The evidence suggests that the mechanism of inactivation involves proteolytic cleavage.

The levels of inactivating activity were low during the experimental growth phase of the yeast culture, but reached high levels in early stationary phase (Fig. 11). A protein inhibitor of the inactivating enzyme, which prevents inactivation of tryptophan synthetase, was also partially purified from the yeast extracts. It is interesting to note the simultaneous appearance of the inhibitor and the protease during the growth of the yeast (Fig. 13).

Recently, Li and Knox (14) obtained evidence suggesting the existence of an inactivating system for tryptophan oxygenase in rat liver.

TABLE X

COMPARISON OF GROUP-SPECIFIC PROTEASE IN VARIOUS ORGANS

Properties	Skeletal Muscle	Intestine		Liver
		Muscle Layer	Mucous Layer	
Common properties				
1. Substrate specificity for pyridoxal enzymes	High	High	High	High
2. Susceptibility to trypsin substrate (TAME)	0	0	0	0
3. Susceptibility to chymotrypsin substrate (ATEE)	Weak	Very strong	Weak	Weak
4. Inhibition by synthetic chymotrypsin inhibitor	No	No	No	
5. Optimum pH in alkaline	9.0	9.0	8.6	8.6
6. Protein nature inhibitor	Exists	Exists	Exists	Exists
7. Coenzyme protection	Exists	Exists	Exists	Exists
8. Influence by diet control	Exists	Exists	No	Exists
9. SH inhibitor	No effect	No effect	No effect	No effect
10. Histidine modification (photo-oxidation)		No activity		
11. Modification of seryl-OH by DFP	Inhibited	Inhibited	Inhibited	Inhibited
Different properties				
1. Molecular weight	12,000–14,000	25,000	30,000	14,000
2. Elution from DE-52 column	0.05 M	0.005 M	0.25 M	0.75 M (after protamine treatment 0.1 M)
3. Reaction with antibody for GSP from intestine (Muscle Layer)	—	+	—	—
4. Susceptibility to ATEE[a]	30%	1000%	25%	20%
5. Susceptibility to TEE[a]	0	100%	0	0
6. Catalytic speed	Very fast	Very slow	Very fast	Fast
7. Effect by Ca^{2+}	Inhibited	No effect	No effect	No effect

[a] 100% represent OTA as a substrate.

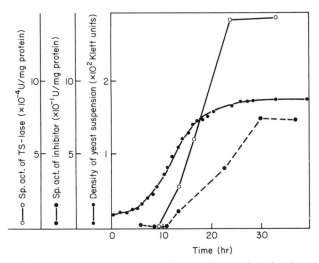

FIG. 13. Dependence of the tryptophan-synthetase-inactivating activity (TS-Iase) on the growth of a yeast culture. The tryptophan-synthetase-deficient mutant DT 29 was grown in "minimal medium" with 0.008% L-tryptophan. From Katsunuma *et al.* (*5*).

VI. Mode of Proteolytic Degradation Catalyzed by Group-Specific Proteases

The fact that the purified proteases show relative specificity for pyridoxal enzyme group, and that their activity is prevented in the holoenzyme, suggests that they attack selected sites which bind coenzymes on the enzyme protein. A result illustrated in Fig. 14 is supported by the results of studies on the mode of degradation of apo-ornithine transaminase by the purified enzyme from skeletal muscle. A reciprocal relationship between the decrease in transaminase activity and increase in acid-soluble, ninhydrin-reactive material was observed. On the other hand, there was no detectable decrease in the total amount of transaminase protein precipitated by 5% tricloroacetic acid (final), even after the transaminase activity was lost completely. The quantity of ninhydrin-positive substance liberated was very small, and only a few peptides were released, as detected by paper chromatography.

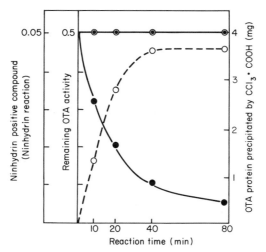

FIG. 14. Relationship between inactivation of ornithine transaminase (OTA) and protein degradation by group-specific protease from muscle. Each reaction mixtures contained 4 mg of crystalline apo-ornithine transaminase, 20 units of group-specific protease, and potassium phosphate buffer, pH 8.5, in a final volume of 1.0 ml. After incubation for 0, 10, 20, 40, and 80 minutes, the remaining transaminase activity (●——●), CCl₃COOH-precipitable protein (⊙——⊙), and ninhydrin-positive substances (○---○) in the CCl₃-COOH-soluble fraction were determined.

These results suggest that the inactivating reaction is associated with very limited proteolytic modification. In confirmation, the inactivated reaction products analyzed by Sephadex column filtration was found to be similar in molecular weight to the original enzyme protein, and one kind of polypeptide was liberated. Analyses of the COOH- and NH₂-terminal sequence of these products are required to establish the nature of the proteolytic cleavage. This may be found to differ depending on the structure of each enzyme, but it may be related to the coenzyme-binding site. Our hypothesis for the intracellular degradation is summarized in Fig. 15.

The enzymes described here may be considered to represent a new types of intracellular protease. We have detected at least four new classes of specific protease isozymes that act only to pyridoxal enzyme group in mammalian tissues, and we have suggested that these may be the initiating factors for degradation of enzymes. We expect

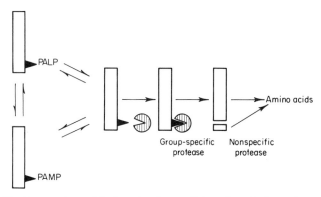

FIG. 15. Proposed model for the intracellular degradation of pyridoxal enzymes.

that the list of such specific inactivating enzymes will continue to grow as other systems are examined.

ACKNOWLEDGMENTS

I would like to thank Dr. Y. Sanada, Dr. E. Kominami, and Mrs. E. Inai for their help in the preparation of the manuscript.

REFERENCES

1. Arias, I. M., Doyle, D., and Schimke, R. T., *J. Biol. Chem.* **244**, 3303 (1969).
2. de Duve, C., and Baudhuin, P., *Physiol. Rev.* **46**, 323 (1966).
3. de Duve, C., and Wattiaux, R., *Annu. Rev. Physiol.* **28**, 435 (1966).
4. Greengard, O., *Advan. Enzyme Regul.* **2**, 277–288 (1964).
5. Katsunuma, T., Schott, S. E., and Holzer, H., *Eur. J. Biochem.* **27**, 520 (1972).
6. Katunuma, N., Katsunuma, T., Kominami, E., Suzuki, K., Hamaguchi, Y., Chichibu, K., and Kobayashi, K., *Advan. Enzyme Regul.* **11**, 37–51 (1973).
7. Katunuma, N., Kominami, E., Kominami, S., and Kito, K., *Advan. Enzyme Regul.* **10**, 289–306 (1972).
8. Katunuma, N., Kominami, E., Kominami, S., Kito, K., and Matsuzawa, T., *in* "Metabolic Interconversion of Enzymes" (O. Wieland, E. Helmreich, and H. Holzer, eds.), pp. 165–177. Springer-Verlag, Berlin and New York, 1972.
9. Kenney, F. T., Reel, J. R., Hager, C. B., and Whittiff, L. L., *in* "Regulatory Mechanisms for Protein Synthesis in Mammalian Cells" (A. San Pietro,

M. R. Lamborg, and F. T. Kenney, eds.), p. 119. Academic Press, New York, 1968.
10. Khairallah, E. A., and Pitot, H.C., *in* "Symposium on Pyridoxal Enzymes" (K. Yamada, N. Katunuma, and H. Wada, eds.), pp. 159–165. Maruzen, Tokyo, 1968.
11. Knox, W. E .,and Greengard, O .,*Advan. Enzyme Regul.* **3,** 247 (1965).
12. Knox, W. E., and Piras, M. M., *J. Biol. Chem.* **242,** 2959 (1967).
13. Kominami, E., Kobayashi, K., Kominami, S., and Katunuma, N., *J. Biol. Chem.* **247,** 6848 (1972).
14. Li, J. B., and Knox, W. E., *J. Biol. Chem.* **247,** 7550 (1972).
15. Manney, T. R., *J. Bacteriol.* **96,** 403 (1968).
16. Schimke, R. T., *J. Biol. Chem.* **239,** 3808 (1964).
17. Schimke, R. T., *in* "Mammalian Protein Metabolism" (H. N. Munro, ed.), Vol. 4, pp. 178–223. Academic Press, New York, 1970.
18. Schimke, R. T., *Curr. Top. Cell. Regul.* **1,** 77–120 (1969).
19. Suzuki, K., Chichibu, K., Katsunuma, T., and Katunuma, N., in preparation (1973).

Author Index

Numbers in parentheses are reference numbers and indicate that an author's work is referred to although his name is not cited in the text. Numbers in italics show the page on which the complete reference is listed.

A

Adams, C. E., 158(131), 165(131), *173*
Adman, N., 78(141), *85*
Agrawal, B. B. L., 63, *83*
Akazawa, T., 4(25, 28, 29), 6(51), 9(51), 12(51), *19, 20*
Albert, A., 166(1), *168*
Alkjaersig, N., 73(132), *85*
Allen, J. M., 30(54), *48*, 150(2), *168*
Allen, M. B., 5(36, 37, 42), *19, 20*
Allison, A. C., 135(3), 158(4), 164, 165, *169*
Alonso, A., 154(20), 158(88), *169, 172*
Altman, R., 62, *83*
Alvarez, B., 158(88), *172*
Ames, S. B., 59(53), *82*
Anderson, C. M. A., 38(14), *46*
Anderson, D., 38(1), *46*
André, J., 136, *169, 170*
Andrews, T. J., 9, *20*, 31(3, 46), 32, 33, 34(2), 35(3, 46), 44(3), *46, 48*
Aoki, A., 133(17), 150(36), *169, 170*
Arbes-Navarrette, I., 149(118), *172*
Arias, I. M., 176(1), 181(1), *202*
Arnon, D. I., 5(36, 37, 41, 42, 43, 44, 53, 62), 7(53), 9(53), 15(62), *19, 20*
Arzadon, L., 66(86), *83*
Ash, B. J., 59(54, 59), *82*
Ashwell, G., 78(145), *85*
Astrup, T., 79, *85*
Atkinson, D. E., 9(56), *20*
Atkinson, M. R., 103(1), *122*

B

Baker, C. A., 166(96), *172*
Baldry, C. W., 5(39), *19*
Baldwin, R. L., 23(37), *47*
Ball, A. P., 73(131), *85*
Bangham, A. D., 76(136), *85*
Barquet, J., 158(88), *172*
Barton, K., 103(2), *123*
Barton, P. G., 57(51), 68, 69(100), 70(104), *82, 83, 84*
Baseman, J. B., 100(35), 106, 114(35), *123, 124*
Bassham, J. A., 2(2, 3), 5(40), 6(40, 49, 50), 12(58), 14(60), 15(49, 50), *18, 19, 20*, 23(5, 16, 74), 30(55), 33 (5), 35(6, 9), 37(9), 38(4), 42(16), *46, 47, 48, 49*
Battellino, L. J., 150(6), *169*
Baudhuin, P., 176(2), *202*
Becker, J. D., 45(7), *46*
Beer, H. P., 120(4), *123*
Benson, A. A., 1, 2, 3(5), *18*, 23(5, 8, 16, 74), 33(5), 35(9), 37(9), 42(16), *46, 47, 49*
Bermingham, M., 27(91), *49*
Bernabe, P., 66(86), *83*
Berre, A., 53(12), 62(12), 63(12), *81*
Berswordt-Wallrabe, R. von, 166(7), 167, *169*
Berthold, A. A., 130(8), *169*
Bidwell, R. G. S., 29(23), *47*
Biggs, M. L., 15(61), *20*
Biggs, R., 56(31), 57(49, 50), 58(49), 59, 78(141), *81, 82, 85*
Bird, I. F., 28(10), *46*

205

Smith, M., 154, *172*
Smith, P. E., 138, 166, *173*
Smith, R., 76(134), *85*
Smulson, M. E., 100(103), 114, 122, *126*
Smyrniotis, P. X., 2(6, 10), 3(6), *18*
Soloforosky, M., 105(72), *125*
Somer, J. B., 57(47), *82*
Soulier, J. P., 56(37), *82*
Spaet, T. H., 61, 78(144), *82, 85*
Speer, R. J., 53(16), *81*
Sperti, S., 97(74), 99(105), 100(73, 104), 102(104), *125, 126*
Spiro, R. G., 103(106), 104(106), *126*
Srivastava, P. N., 158(61, 131), 165(131), *171, 173*
Stadtman, E. R., 9(57), *20*
Stafford, H. A., 40(76), *49*
Stambaugh, R., 158(132), 165(132), *173*
Starks, M., 30(54), *48*
Steadman, R. A., 166(140), *173*
Steele, B. B., 76(135), *85*
Steinbeck, H., 166(7), 167, *169*
Steinbeck, R., 97(62), *125*
Steinberger, A., 166(135), 168(135), *173*
Steinberger, E., 146(133), 158(134), 160, 161(23), 166, 167(23), 168(133, 135), *169, 173*
Stepka, W., 2(3), *18*, 23(16), 42(16), *47*
Stern, H., 155, 156, *171, 173*
Steward, J., 56(31), *81*
Stiller, M., 2(12), *19*
Stoll, P. J., 78, *85*
Storosser, M. T., 89(13), 114(13), 115(13), 116(13), 120(13), *123*
Strauch, S., 153(109), 166(109), *172*
Strauss, N., 90, 105(107), *126*
Sugano, H., 143(144), *173*
Sugimura, T., 89(25, 108, 109), 114(25, 45, 98, 109), 115(45, 97, 108, 110, 111), 116(25, 86, 98), 117(86), 118(26, 99), 119(28, 29, 119), 120(25,

45, 67, 68, 70, 109, 110), 121(75), 122(108), *123, 124, 125, 126, 127*
Sugiyama, T., 4(25, 28, 29), 6(51), 9, 12(51), *19, 20*
Summaria, L., 66(86), *83*
Sung, M. T., 158(138), *173*
Sung, S. C., *127*
Surgenor, D. M., 76(135), *85*
Susumu, I., 140(46), *170*
Suzuki, K., 177(6), 181, 188(6), 189(6), *202, 203*
Swart, A. C. W., 69(101), *84*
Szego, C. M., 166, *173*

T

Tabita, F. R., 14(59), *20*
Takagi, T., 53(17), *81*
Takeda, M., 88(82), 114(82), 115(82), 116(82), 117(82), 118(82), *126*
Taketa, K., 15(64), *20*
Tanabe, T., 101(53), 104(53), 105(53), *124*
Tanford, C., 76(134), *85*
Tanner, H. A., 30(77), 42(77), *49*
Tarnoff, J., 153(73), 166(73), *171*
Taylor, F. B., 79(151), *86*
Telfer, A. R., 39(78), *49*
Teller, D. C., 53(9), 57(9), 77(9), *81*
Tepperman, J., 137, *173*
Theodor, I., 69(94), 70(94), *84*
Thompson, A. R., 57(11), 59(58), 60, *81, 82*
Thompson, D. W., 129(142), *173*
Thompson, E. O. P., 4(26), *19*
Titani, K., 65(82), 66, 67, *83*
Tolbert, N. E., 22(80, 81, 82), 23(15, 16, 31, 32, 37, 62, 79), 24(80), 25(31, 39, 40, 41, 80), 26(79, 80), 27(34, 82), 29(52, 84), 30(33, 73), 31(3, 46), 32(2, 3, 46), 33(2, 3, 46), 34(2), 35(3, 9, 38, 46), 36(63, 64, 80), 37(9, 31), 40, 41(66, 73), 42, 43(38, 48, 82, 84), 44(3, 65), 45(33), *46, 47, 48, 49*
Tove, S. B., 25(98), *49*

Subject Index